Biochemical studies on Sialic acid and Lectin in sera of patients with Multiple Myeloma

Prof . Dr.Sami AL-Mudhaffar

Dr.Hasanain Kamil AL-Bermani

Abbreviations

APP	Acute phase proteins
B	Bound
B_{max}	Maximal binding capacity
BSA	Bovine serum albumin
$CA_{19\text{-}9}$	Carbohydrate antigen 19-9
CEA	Carcinoembryonic antigen
CMP	Citidine monophosphate
EDTA	Ethylene diamine tetra acetic acid
F	Free
HPLC	High performance liquid chromatography
Ig	Immunoglobulin
Ka	Affinity constant
Kd	Equilibrium dissociation constant
LBSA	Lipid-bound sialic acid
MM	Multiple myeloma
MP	Mucoid proteins
μU	Microunit
NANA	N-Acetylneuraminic acid
NBT	Nitro blue tetrazolium
NeuGc	N-Glycolylneuraminic acid
NSB%	Percent of non-specific binding
SB%	Percent of specific binding
SOD	Superoxide dismutase
TB%	Percent of total binding
TP	Total protein
TSA	Total sialic acid

7

Table Of Contents

List of Figures

List Of Tables

Introduction

&

Literature Survey

INTRODUCTION

The term sialic acid is used to describe derivatives of neuraminic acid. There are more than twenty natural derivatives of neuraminic acid. The most commonly occurring of these is N-acetylneuraminic acid (NANA), also N-glycolylneuraminic acid (NeuGc) is an important derivative of neuraminic acid contains a glycolyl group instead of an acetyl substituent. Other derivatives contain one or more O-acetyl substitutions on NANA molecule.

Purified sialic acids are colorless, they do not melt but decompose with discoloration. The infrared spectra account for all functional groups known from chemical data to be present in sialic acids.

Sialic acids are soluble in water, insoluble in organic solvents. They are very unstable to both acid and alkali sialic acids have a great tendency to form esters.

N-Acetylneuraminic acid was synthesized in 8-10% yield from the condensation of N-acetyl-D-mannosamine and oxaloacetic acid at pH 11. Several other methods to synthesize NANA have also been reported.

Sialic acids have a restricted occurrence in nature, being present in mammalians and some bacteria but not in plants. They are encountered in glycosidic linkages at the nonreducing end of a variety of biooligomers.

Of the enzymes acting on sialic acids, neuraminidase, this enzyme catalyzes the enzymatic release of sialic acid from a wide range of glycoconjugate substrates. Sialyltransferase catalyzes the transfer of sialic acid from CMP-sialic acid to appropriate acceptors on oligosaccharide chains in the process of biosynthesis of glycoproteins.

A variety of methods are available for the measurement of both total and lipid-bound sialic acid in serum or plasma. These including the calorimetric procedures, fluorometric procedures, enzymatic procedures, and high performance liquid chromatographic (HPLC) procedures. The newer HPLC procedures can detect picogram levels of sialic acid and are relatively free of interference seen with classical procedures.

The negative charge on sialic acid owing to its carboxyl group, enables it to play a role in cellular functions, such as transport of positively charged compounds, cell–to–cell repulsion, influencing conformation of glycoprotein on cell membranes, and even masking the antigenic determinants on receptor molecules.

Focus on sialic acid as a tumor marker should be examined from the perspective of aberrant glycosylation in cancer cell membranes owing to activation of new glycosyltransferases that are characteristic of tumor cell. Serum sialic acid has been reported to be elevated in different cancers, and has been used as a tumor marker for these cancers.

The role played by sialic acid in tumor cell metastasis including increased capacity to adhere to vascular endothelium, and decreased capacity of cancer cell to be destroyed by host defence mechanism.

The high sensitivity of sialic acid as a tumor marker has been reported in a variety of cancerous conditions. Its specificity is relatively low since there is also an increase in sialic acid – rich glycoproteins in inflammatory diseases. Sialic acid measurements, however, have value in monitoring cancer patients during treatment.

Glycoproteins, which contain carbohydrate groups attached covalently to the polypeptide chain, represent a large group of wide distribution and considerable biological significance. Glycoproteins having a very high content of carbohydrate are called proteoglycans.

In vertebrates, most of the glycoproteins are extracellular in occurrence and function, among these are the cell-coat glycoproteins, blood glycoproteins, and the antibodies. In membrane glycoproteins, sugars are attached either to the amide nitrogen atom (N-linkage) or to the oxygen atom (O-linkage).

Changes in normal serum glycoproteins have long been associated with malignancies, Elevated levels of glycoproteins have been reported in different cancers.

Superoxide dismutase presents in all oxygen-metabolizing cells, its physiological role is. to protect these cells from damage by superoxide free radical, since it catalyzes the breakdown of this radical. Four different forms of SOD have been found, one of these contains copper and zinc, two kinds contain manganese, and the fourth type contains iron. A decrease in the SOD activity has been reported in all patients with cancer diseases.

Lectins are multivalent carbohydrate- binding proteins with the ability to agglutinate erythrocytes, bacteria, and other cells. Over 50 lectins have been purified from the seeds of plants, relatively few sialic acid-specific lectins have been identified among which only one is commercially available. In vertebrates, lectins are subdivided into two kinds, membrane lectins which require detergents for their extraction, and soluble lectins. Only few of invertebrate lectins have been purified, most of these are specific for sialic acid.

Plant lectins are involved in the fixation of nitrogen to legumes and protection against phytopathogens, animal lectins are involved in the endocytosis of glycoproteins, regulation of cell migration and adhesion, and binding of bacteria to epithelial cells. Lectins have been used as useful tool for blood typing and purification of glycoconjugates.

Multiple myeloma is a malignant proliferative disease of immunoglobulin-producing cells. The clinical features include backache, pathological fractures, anaemia, repeated infections, hypercalcaemia, polyuria and others. The diagnosis of multiple myeloma depends on three principal findings, lytic bone lesions, plasma cell accumualtion in the bone marrow, and the presence of monoclonal proteins in serum and urine. Chemotherapy and radiotherapy are commonly used in the treatment of multiple myeloma.

1.1 SIALIC ACID

General Knowledge

Sialic acid is a generic name given to a family of N- and O-substituted derivatives of neuraminic acid, a nine-carbon polyhydroxyaminoketo acid sugar(5-amino-3,5-dideoxy-D-glycero-D-galactononulosonic acid) are widely distributed in nature [1].

Neuraminic acid has five hydroxyl groups at the C-2, C-4, C-7, C-8, and C-9 positions in its molecule. The most reactive position to acetylation is C-9 and the next most reactive position is C-4. For selective acetylation of other positions the more reactive hydroxyl groups must be protected [2].

There are more than twenty natural derivatives of neuraminic acid. In the structure of neuraminic acid depicted in figure (1-1), the hydroxyl groups could either be methylated or esterified with acetyl, lactyl, phosphate, or sulfate groups. The amino group is substituted by an acetyl group thus forming N-acetylneuraminic acid. Alternatively, the amino group could also be substituted by a glycolyl residue [3].

In animals, sialic acids are found in α-glycosidic linkage as the outermost sugar of glycoproteins and glycolipids, except in the case of the polysialolyl linkage, and certain other rare exceptions [4]. Sialic acids have also been found in prokaryotic cells as a constituent capsular polysaccharide of a few genera of pathogenic bacteria [5].

Figure (1-1)

R_1 = H. Acetyl (4, 7, 8, 9). Lactyl (9), Methyl (8). or sulfate (8).

R_2 = Acetyl or Glycolyl

R_3 = Gal., GalNAc, GlcNAc or SA.

 The sialic acids: The parent molecule neuraminic acid is shown in partially stylized form in the chair conformation. The individual carbon atoms are numbered from 1 through 9. R_1 indicates possible glycosidic linkages. R_2 indicates substitutions of the N group. and R_3 indicates O-substitutions. The known types of substituent are indicated on the figure for each case.

1.2 SIALIC ACID DERIVATIVES

 Over 20 different naturally occurring sialic acids have been reported in glycoconjugates of various animal species [6].

1. N-acetylneuraminic acid(NANA).

 N-acetylneuraminic acid is the most commonly occurring one of sialic acids. It is a nine-carbon acidic sugar, which is an important constituent of the oligosaccharide chains found in the glycoproteins and glycolipids of the cell coats and membranes of animal tissue [7]. It was first encountered by Gottschalk upon examination of the action of influenza viruses on various mucins and later by Klenk and co-workers upon acidic hydrolysis of mucous substances[8].

2. N-GLYCOLYLNEURAMINIC ACID (NeuGc)

N-glycolylneuraminic acid differs by a glycolyl instead of an acetyl substituent at the C-5 amino group[9]. Most of the other sialic acid derivatives contain one or more O-acetyl substitutions of the hydroxyl groups at C-4, C-7, C-8, and C-9 [9], such as 4-O-acetyl-N-acetyl neuraminic acid and 9-O-acetyl-N-acetyl neuraminic acid.

1.3 SYNTHESIS OF SIALIC ACID

A) Biosynthesis of sialic acid

The following sequence of reactions shows the formation of N-acetylneuraminic acid [10].

acylglucosamine-2-epimerase

UDP-N-acetylglucosamine + H_2O ⟶ N-acetylmannosamine + UDP

N-acetylmannosamine Kinase

N-acetylmannosamine +ATP ⟶ N-acetylmannosamine 6-phosphate +ADP

N-acetylneuraminate-9- phosphate synthase

N-acetylmannosamine-6-phosphate + phosphoenolpyruvate ⟶

N-acetylneuraminic acid –9- phosphate +Pi

N-acetylneuraminate-9- phosphatase

N-acetylneuraminic acid –9- phosphate + H_2O ⟶

N-acetylneuraminic acid +Pi

B) Chemical synthesis of sialic acid

The ever increasing interest in N-acetylneuraminic acid prompts to evaluate the available synthetic methods for large scale preparation.

The synthesis of NANA from N-acetyl-D-glucosamine and oxaloacetic acid at pH 11 in 1-2 % yield has been reported by Cornforth et al [11].

After the finding[12] that NANA is structurally related to N-acetyl-D-mannosamine and not to N-acetyl-D-glucosamine, Carroll and Cornforth [13] sought to improve this synthesis by employing N-acetyl-D- mannosamine in the condensation reaction. The later authors thus claimed an 8-10% yield of NANA.

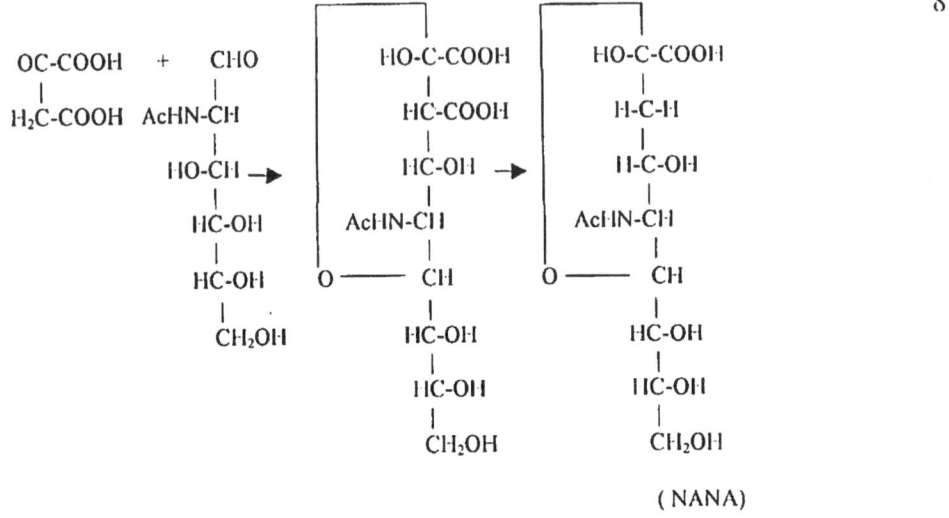

(NANA)

N-acetylneuraminic acid has been used as a starting material for the synthesis of other sialic acid derivatives. Different groups have been substituted on NANA molecule with selective protection of the other positions[14].

1.4 OCCURRENCE OF SIALIC ACIDS IN TISSUES AND BODY FLUIDS

Sialic acids are present in higher animals and certain bacteria but not in plants or lower invertebrates [15]. Within mammalian species the occurrence of sialic acids varies considerably[6].

In human plasma a large quantity of sialic acid is found in orsomucoid, alpha-1-antitrypsin, haptoglobin, ceruloplasmin, fibrinogen, complement proteins, and transferrin [16].

Also, it was purified from cervical mucus, ovarian cyst, pseudomyxomatous gel, lipid-free fraction of brain, and liver [17].

Of sialic acids found in submaxillary mucins, N-acetylneuraminic acid is the predominant sialic acid in monkeys, and N-glycolylneuraminic acid is the predominant one in swine, whereas the submaxillary mucin in sheep contains N-acetylneuraminic acid with an admixture of only 1-3% N-glycolylneuraminic acid. Also it has been found that 9-O-acetyl-N-acetylneuraminic acid and 4-O-acetyl-N-acetylneuraminic acid are major sialic acids in mucins of bovine and equine species respectively[18].

1.5 METHODS FOR DETERMINATION OF SIALIC ACIDS

A variety of procedures have been used for the measurement of total sialic acid. These can be broadly classified as colorimetric, fluorometric, enzymatic and highly sensitive high performance liquid chromatographic (HPLC) procedures.

1.5.1 COLORIMETRIC PROCEDURES

Two classical procedures have stood the test of time. One uses resorcinol and the other uses periodic and thiobarbituric acids.

1.5.1.1 RESORCINOL METHOD

The resorcinol based assay uses heat and strong acid to hydrolyze glycosidic bonds. The released free sialic acids are reacted with resorcinol and copper ions to give colored compound which is extracted and measured at 580 nm.

To overcome interference from sugars forming furfural and furfurol analogues, such as pentoses, and other interferents, such as glucuronic acid and 2-deoxyglucose which incidentally have an absorption maxima at 450nm. and second maxima at 580nm. measurements are also taken at 450nm. sialic acid concentration is then calculated using simultaneous equation.[19]

1.5.1.2 THE THIOBARBITURIC ACID ASSAY

It measures only free sialic acid that is released after an initial hydrolysis step. In this procedure formyl pyruvic acid formed as a result of periodic acid oxidation of free sialic acid is reacted with thiobarbituric acid to yield a red color which is measured at 549nm. Interference owing to malonaldehyde, the oxidation product of 2-deoxyglucose, is corrected by taking measurements also at 532nm.[20]

1.5.2 FLUOROMETRIC PROCEDURES

In a typical and more specific assay formaldehyde that is formed upon oxidation of free sialic acid by periodic acid is reacted with acetyl acetone. The yellow product is excited at 410nm, and the resulting fluorescence is measured at 510nm.[21]

1.5.3 ENZYMATIC PROCEDURES

Enzymatic assays are based on conversion of free sialic acids released by the enzyme neuraminidase to pyruvate and N-acetylmannosamine with the aid of the enzyme acetyl neuraminic acid pyruvate lyase or NANA-aldolase. The resulting pyruvate can be coupled to the lactate dehydrogenase NADH system to measure the oxidation of NADH to NAD at 340nm. Alternatively, pyruvate can be coupled to pyruvate oxidase, flavine adenine dinucleotide (FAD), and thiamine pyrophosphate (TPP)to form hydrogen peroxide, which in turn is coupled to peroxidase in presence of 4-aminoantipyrine and a toluidine derivative to form a red chromogen which is measured at 550nm[22].

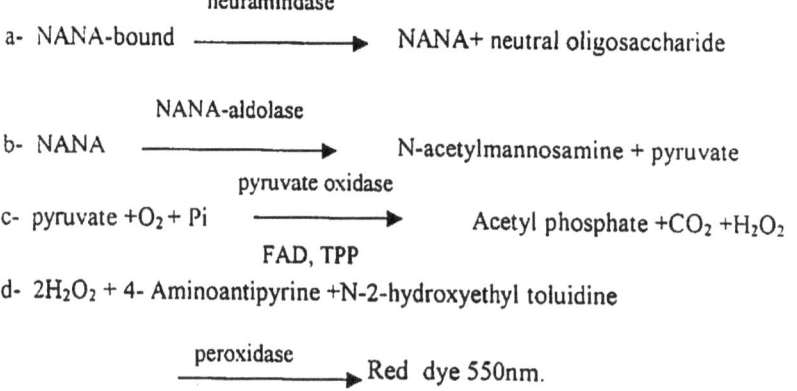

neuramindase

a- NANA-bound ⟶ NANA+ neutral oligosaccharide

NANA-aldolase

b- NANA ⟶ N-acetylmannosamine + pyruvate

pyruvate oxidase

c- pyruvate $+O_2+$ Pi ⟶ Acetyl phosphate $+CO_2 +H_2O_2$

FAD, TPP

d- $2H_2O_2$ + 4- Aminoantipyrine +N-2-hydroxyethyl toluidine

peroxidase

⟶ Red dye 550nm.

Reactions involved in atypical enzymatic assay for the measurement of sialic acid

1.5.4 HIGH PERFORMANCE LIQUID CHROMATOGRAPHIC PROCEDURES

The HPLC procedures provide the ultimate sensitivity. In one such procedure, sialic acid released from the sample by acid hydrolysis is converted to highly fluorescent derivatives by reacting with a fluorogenic agent for alpha-keto acids such as 1,2-diamino 4,5-methylenedioxy benzene in dilute sulfuric acid. The fluorescent derivatives are separated on an octadecyl (C18) bonded silica column using a reverse phase solvent system. The chromatographic step taken only 12 minutes allowing detection of levels as low as 7.7 pico grams (pg)of N-acetylneuraminic acid and 7.5 pg of N-glycolylneuraminic acid, in an injection volume as small as 10 micro liters. The procedure is capable of analyzing precisely sialic acid in a 5 μL of serum sample[23].

1.5.5 MEASUREMENT OF LIPID – BOUND SIALIC ACID (LBSA)

The basis for the measurement of LBSA in serum is the increased levels of gangliosides seen in patients in various types of cancers. Apparently, these gangliosides are shed from the tumor cell surface; since they have a relatively long half-life compared to lipids lacking sialic acid, their accumulation in serum lends itself to its measurement [24].

However, there is a difference in levels of LBSA depending on the procedure used for measurement. This is because some procedures measure predomenantly glycoprotein-bound sialic acid, thus grossly overestimating the LBSA. Essentially, these procedures involve extraction of glycolipid-bound sialic acid with solvents such as chloroform-methanol and the gangliosides are seperated from other lipids either by phase partition or precipitated with a one percent phosphotungstic acid solution. The isolated LBSA fraction is measured by either the classical resorcinol or periodic –thiobarbituric acid method[25].

1.6 BIOLOGICAL FUNCTION OF SIALIC ACID

1.6.1 LOCATION OF SIALIC ACID

Sialic acids are present in high concentrations as a terminal component at the non-reducing end of carbohydrate chains of glycoproteins and glycolipids such as gangliosides, or polysaccharides.

Sialic acids are usually located on the outer cell membrane linked to other sugars such as galactose or N-acetylgalactosamine by glycosidic bonds. While the usual location of sialic acids is at the terminal position of oligosaccharides, they may also be located in the side chain of oligosaccharide units in gangliosides and some glycoproteins.

The fact that the carboxyl group of sialic acid is always free, and because its pka of 2.6 is negatively charged at physiological PH, together with its location as a terminal non-reducing unit of oligosaccharides, enables it to have a marked influence on the physical and chemical properties of glycoproteins[3].

1.6.2 PHYSIOLOGICAL ROLE OF SIALIC ACID

The elucidation of the roles of sialic acids in moderating a range of biological properties and functions is an increasingly important area of biochemical research[26].

Sialic acids have a number of major biological functions:-

1- The human red blood cell is studded with nearly 20 million molecules of sialic acid on the outer cell membrane which contributes to its electronegative charge (zeta potential)[3], sialic acid has been shown to be the main carrier of surface negative charge in most mammalian cells[27].

2- Its large electronegative charge, with a pka value of around 2 under physiological conditions, has implications for the cell-cell repulsion due to an electrostatic barrier and by this repulsion prevents red blood cells from aggregating.[28].

3- Owing to its negative charge, sialic acid can bind positively charged molecules and thus play a role in the transport of such molecules.[3].

4- Sialic acids are found as essential component of many cell surface receptors for a wide range of endogenous (interferon, peptide hormones, and certain tumor-specific and blood group-specific antibodies)and exogenous (animal viruses, mycoplasma, plant and animal lectins, bacterial toxins) substances, their presence is both a blessing as well as a scourge. Thus they have a role in the cellular actions of hormones, such as insulin, and also can modulate aminoacid transport in some cells. [3]. On the other hand, infection by bacteria or virus is solely dependent on the presence of sialic acid as a component of specific receptors for the microorganisms on the cell membrane[29].

5- Sialic acids are often part of antigenic determinants of glycolipids or glycoproteins, thus sialic acid molecules contribute to the specificity of blood group substances[28].

6- The negative charge on sialic acid have an influence on the conformational states in cell membranes, the expression of enzymatic activity of glycoprotein enzyme,and even resistance to proteolytic enzyme degradation[28].

7- The clustering of cell membrane glycoproteins owing partly to the repulsion of their oligosaccharide sialic acid residues, is important for cell rigidity, since the loss of sialic acid molecules can increase the deformability of cells [28]. So sialic acid is involved in maintaining the shape of human erythrocytes[30].

8- An intriguing role for sialic acid is in its ability to act as biological masks by preventing ligands from recognizing receptors. Thus, a glycoprotein layer rich in sialic acid acts as an immune barrier between mother and fetus. Indeed this masking "anti-recongnition" effect is lost by the removal of terminal sialic acid residues from oligosaccharide chains since it leads to the exposure of a penultimate galactose residue which is then recognized and bound by naturally occurring antibodies, thus facilitating the removal of glycoprotein or the cell by the reticuloendothelial system[3].

9- Sialic acids are thought to be important in determining the surface properties of cell and has been implicated in determining cellular adhesivness, the ability of intravenously injected cells to implant in various organs, cellular motility and invasivness in vitro and immunogenicity[31].

10- The O-substituted sialic acids can influence enzymatic reactions in the catabolism of glycoconjugates, slow down or inhibit bacterial and viral sialidases, and alter immunopotency of the sialoconjuagates[32].

I.7 ALTERATION OF CARBOHYDRATE CHAINS IN MALIGNANCY

During neoplastic transformation, the carbohydrate chains in glycolipids and glycoproteins are frequently altered.

There is a close relationship between the expression of certain carbohydrate antigens and oncogenesis [33]. In an elegant study examining the significance of the linkage of sialic acid residues in cancer-associated carbohydrate antigens, by using specific monoclonal antibodies, it was demonstrated that not all sialic acids are specific to cancer. Indeed, this study demonstrated that there are significant variations in the cancer specificity depending on the difference in linkage in sialic acid residues. Thus for pancreas, the 2-3sialylation at the terminal galactose of the Lewis[a] (Le[a]) antigen appeared to be significantly increased in the cancer cell in contrast to the decreased 2-6 sialylation at the penultimate N-acetylglucosamine during the course of development of tumor[34].

Indeed, the epitope of carbohydrate antigen CA19-9, well known as a tumor marker for pancreatic and gastric cancers, is sialylated lacto-N-fucopentaose II which is synonymous with the Lewis[a] antigen. Interestingly, an antigen that has a chemical structure isomeric to the Le[a] antigen is the sialylated Le[x] antigen, which is a tumor associated marker for lung adenocarcinomas[33].

1.8 SIALIC ACID AS A TUMOR MARKER

1.8.1 SIALIC ACID AND TUMOR

There has been much discussion of the relation between sialic acid and cancer. The relevance of sialic acid to the tumor cell is apparent from the increased sialylation of tumor cell surface and siglytransferase activity observed in many cancer cells[29].

The aberrant glycosylation found in cancer cell membranes is presumably due to the activation of new glycosyltransferases that are characteristic of tumor cells and are absent or present only in small quantities in normal cells [35]. Thus, for instance, a relatively specific sialyltransferase is found to be present by as much as 2.5 to 11 times in greater amounts in transformed cells when compared to normal cells.

Higher amounts of sialic acid were detected in cancer patients than in normal healthy donors and patients with non-cancer diseases.[36,37].

Elevations of serum/ plasma sialic acid as reflected by N-acetylneuraminic acid have been reported in patients with various types of cancers including colon[38], lung[39], bladder[40], ovarian [41], cervical cancer[42], tumors of central nervous system [43], and malignant melanomas[44].

Serum/plasma sialic acid levels have been correlated with the stage of disease [45], tumor burden [46], degree of metastasis[47], prognosis[48], and detection of early recurrence[49].

Also,the elevations of sialic acid concentrations may indicate resistance of the cancer to previously given chemotherapy[3].

1.8.2 SIALIC ACID AND TUMOR CELL METASTASIS

Metastasis is a selective process by which certain tumor cells disseminate to form secondary foci at distant sites[50].

One could theoretically envisage three steps in the metastatic process that involve cell surface carbohydrate interactions [51]:

a- in the release of tumor cells from the primary tumor mass due to altered homotypic adhesion phenomena.

b- in the mechanism of blood transportation of metastatic tumor cells by heterotypic cell interactions.

c- in the arrest in organs by specific interactions with the target tissue[51].

Alterations in the level of membrane-bound sialic acid or a rearrangement of sialic acid in the plane of the membrane could be responsible for some of the altered properties of transformed cells[52].

It has been suggested that the ability of murine tumor cells to metastasize spontaneously from subcutaneous sites is positively correlated with the total sialic acid content of the cells in culture. Other factors that influence metastasis include the extent to which sialic acid molecules are exposed on the tumor cell surface and most strongly to the degree of sialylation of galactosyl and N-acetylgalactosaminyl residues present on the cell surface oligosaccharide chains[53].

Apparently, the tumor cells use their heavily sialylated surfaces as a mask to evade recognition by immune surveillance, and this facilitated the metastatic spread[29].

The increased amount of sialic acid on the tumor cell surface can, by increasing adhesiveness, contribute to the formation of larger tumor emboli.

Metastatic spread is also facilitated by sialic acid molecules increasing the adherence of tumor cells to vascular endothelium and by increasing the ability to aggregate platelets[53].

1.8.3 USEFULNESS OF TSA & LBSA FOR MONITORING

In a typical study, four groups of cancer patients were evaluated these groups included 69 patients with bladder cancer, 58 patients with lung cancer, 31 patients with cancer of the uterus, and 29 patients with breast cancer. In addition to TSA and LBSA levels in sera, the carcinoembryonic antigen (CEA) levels were also measured. The sensitivity of the three assays prior to initiation of radiotherapy, in relation to established cut-off values given in parenthesis were: TSA 89.3% (80 mg/dL), LBSA 88.8% (20 mg/dL), and CEA 26.8% (5ng/mL). After completion of radiotherapy, the overall response to treatment as judged by percent of patients with a final serum level below the zero time value for each of these markers were: LBSA 85.6%, TSA 81.3% and CEA 65.8%. Thus, in this study the diagnostic sensitivity of LBSA and TSA in terms of their ability to detect true positives was more than three times greater than CEA. For monitoring response to treatment, TSA and LBSA were able to follow correctly 15% more patients with post treatment values below the zero time values when compared to CEA[54].

1.8.4 SIALIC ACID IN SOME CANCERS

Sialic acid bound to membrane glycoproteins and glycolipids apparently enters the circulation by either shedding or by cell lysis.

Approximately 98 to 99.5 percent of total sialic acid found in serum or plasma is bound to glycoproteins. Only a very small fraction of sialic acid is bound to lipids which is mainly in the form of gangliosides.

Normal levels of total sialic acid in serum are approximately in the range of 51 to 84mg/dL. In contrast, the contribution of the pure lipid fraction to the total sialic acid level is barely in the range of 0.4 to 0.9mg/dL[55].

In a recent study on the usefulness of total and lipid-bound sialic acid in lung cancer, 152 patients with lung cancer, 107 patients with benign pulmonary disease and 207 normal controls were studied. The data obtained in the study were 118.7±40.4, 89.7±0.12, and 64.3±8.5 mg/dL respectively. The data of LBSA measurements were 28.6±11.46, 21.5±0.053 and 15.5±3.4 respectively. From these data it is apparent that the mean concentration of TSA and LBSA were significantly higher in lung cancer patients when compared to benign and normal controls. The sensitivity of TSA measurement was 86.5 percent and 77 percent for LBSA. The specificity was 44 percent for both TSA and LBSA[3].

LBSA levels have been reported to be useful in monitoring patients with malignant melanoma. In one study when tumor recurrence was correlated with elevated LBSA, the increased level was found as early as 9.3 months prior to recurrence.

Higher levels of TSA and LBSA have been reported in leukemia patients compared to patients with anemia. The TSA levels were significantly higher in acute myeloid leukemia (AML) compared to chronic myeloid leukemia (CML) and acute lymphatic leukemia (ALL) patients. The LBSA levels were significantly elevated in AML patients as compared to other leukemic patients. The sensitivity of sialic acid as a marker for leukemia is high with the sensitivity of LBSA approaching 85 percent[3].

The utility of sialic acid measurements in colorectal cancer has been examined. In one study it was shown that although TSA and LBSA measurements were not useful for detecting early-stage colorectal cancer, TSA and TSA/Total protein ratio (mg/g) were significantly elevated in each colorectal cancer subgroup compared to values obtained with normal subjects. In another study where 146 colorectal cancer patients were studied prior to surgical therapy, the TSA/TP ratio (mg/g) for colorectal cancer was 13.4 in contrast to 9.7 for normal controls and 12.1 for pathological controls. Thus, the TSA value normalized to total protein (TSA/TP) may have utility in detecting colorectal cancer and following patients on treatment[3].

1.9 GLYCOPROTEINS

Among the several different classes of conjugated proteins, the glycoproteins, which are an important cell surface constituents, are linear polymers of amino acids with branching chains of carbohydrates which include hexoses (galactose and mannose), sialic acid, methylpentose (fucose), and amino sugars (N-acetylglucosamine, and N-acetylgalactosamine[56].

The glycoproteins represent a large group of wide distribution and considerable biological significance. In fact, on closer study many proteins once thought to be simple proteins, i.e. containing only amino acid residues, have been found to contain carbohydrate groups. Table (1.1) shows the biological distribution of some major types of glycoproteins.

The percent by weight of carbohydrate groups in different glycoproteins may vary from less than one percent in ovalbumin to as high as eighty percent in the mucoproteins. Glycoproteins having a very high content of carbohydrate are called proteoglycans[57].

In membrane glycoproteins, sugars are attached either to the amide nitrogen atom in the side chain of asparagine (termed an N-linkage) or to the oxygen atom in the side chain of serine or threonine (termed an O-linkage) [58].

Glycoproteins are found in all forms of life. In vertebrates most but not all of the glycoproteins are extracellular in occurrence and function or are secreted from cells, it has accordingly been suggested that one purpose of the attached sugar residues is to label the protein for export from the cell. Among the glycoproteins having extracellular location or function are the cell-coat glycoproteins, the blood glycoproteins, the circulating forms of some protein hormones, the antibodies, various digestive enzymes secreted into the intestine, the mucoproteins of mucous secretions, and the glycoproteins of extracellular basement membranes[57].

Table(1-1):Some glycoproteins, grouped according to biological occurrence[57]

Blood plasma	Fetuin, α_1-Acid glycoprotein, Fibrinogen Immunoglobulins and blood-group proteins
Urine	Urinary glycoprotein
Hormones	Chorionic gonadotrophin, Follicle-stimulating hormone, and Thyroid-stimulating hormone.
Enzymes	Ribonuclease B, B-Glucuronidase, Pepsin, and serum Cholinesterase.
Mucus secretions	Submaxillary glycoproteins, Gastric glycoproteins.
Connective tissue	Collagen
cell membranes	Glycophorin of erythrocyte membrane.

Glycoproteins and cancer

Many properties of mammalian cells are expressed at, or mediated through, the cell surface. Among these properties are those which distinguish malignant cell from normal cell. .

As neoplastic changes are expressed at the cell surface, altered surface characteristics are essential for the abnormal growth and behavior of malignant cells[59].

Glycoproteins, a major structural component of the cell surface, play an important role in many biological functions such as cell-cell interactions, growth regulation, differentiation, and malignant transformation, and tumor cell metastasis may require alterations in membrane properties determined by cell surface glycoproteins[60].

The measurements of biochemical markers is being increasingly used for early diagnosis and monitoring the progress of cancer. Increased levels of glycoproteins have been shown to occur frequently in patients with cancers[61].

Elevated levels of glycoproteins have been reported in the serum of patients with metastatic breast carcinoma, uterine cervical carcinoma, and lung cancer. These elevations are due to the exponential growth of tumor cells, which results in a rapid rate of membrane glycoprotein turnover and shedding of these excessive glycoproteins into the sera[61].

1.10 SUPEROXIDE DISMUTASE

Superoxide dismutase is believed to be present in all oxygen metabolizng cells but lacking in most obligate anaerobes, presumably because its physiological function is to protect the cell from damage by the highly reactive superoxide free radical (O_2^-) generated by aerobic metabolic reactions[62].

The enzyme catalyzes the breakdown of the superoxide free redical according to the reaction:

$$O_2^- + O_2^- + 2H^- \rightarrow H_2O_2 + O_2$$

Superoxide is formed by the non-electron reduction of oxygen, and has been identified as a product in a number of biological reactions. It is particularly likely to be formed in the red cell and has been shown to be produced when oxyhemoglobin is autoxidized to methemoglobin[63].

$$Hb - Fe^{-2} + O_2 \longrightarrow Hb - Fe^{-3} + O_2^-$$

The red cell is susceptible to damage by superoxide free redical, and if this is allowed to build up because a deficiency in superoxide dismutase, there is precipitation of hemoglobin as Heinz bodies with an associated hemolytic anemia[63].

Superoxide dismutase was first identified and purified from the red cell by McCord and Fridovich[64] who found it to be synonymous with erythrocuprein. The enzyme has since been detected in a large number of tissues and organisms.

Four different forms of SOD have been found. One of these, which is found in cytosol and intermembrane space of mitochondria of eukaryotic cells, contains copper and zinc and is entirely unrelated, except in its activity, to the other three. There are two kinds of SOD that contain manganese. One of these is found in the matrix of mitochondria, and the other is found in the matrix of bacteria such as Escherichia coli. The fourth type of SOD contains iron and has been found in the periplasmic space of E.coli[62].

Yamanaka and Deamer were the first to report that the activity of SOD is abnormal in transformed cells. Decreased SOD activity has been found in all malignant tumors investigated so far[65]. In a recent study, it has been shown 30% lower SOD activity in the sera of children with cancer compared with the sera of healthy controls[66].

1.11 THE LECTINS

Lectins (A lectin from the latin *legere*: to chose) are multivalent specific carbohydrate–binding proteins or glycoproteins of non-immune origin which are widely found in nature including in plants and vertebrate tissues[67,68].

They are grouped together because they have the ability to agglutinate erythrocytes, bacteria, and other normal and malignant cells that display more than one saccharide of sufficient complementarity, and/or precipitate polysaccharides, glycoproteins, and glycolipids[68,69].

Binding of lectins to the carbolydrate moieties of glycoproteins on the surface of cells has been shown to be involved in a variety of biological recognition processes [67].

They were discovered in plants, but are also found in all other categories of living things. To more sharply define this category, it has been proposed that a number of related proteins be excluded. Among these are monovalent carbohydrate- binding toxins such as ricin(because of their valence); carbohydrate –binding immunoglobulins(presumably because they can be included in another structural and functional category); and carbohydrate-binding enzymes (because they are originally not agglutinins)[69].

Over fifty lectins have been purified and are available commercially. They have been purified mainly from the seeds of plants and exhibit a wide variety of specificities [70].

For years they found wide applications as reagents and, little thought was given to the role they play in the plants from which they were derived[71].

In past decade, interest in the function of lectins has increased considerably. This was stimulated by their isolation from many other organisms, including vertebrates and cellular slime molds[71].

1.11.1 SIALIC ACID- BINDING LECTINS

Lectins are classified into a small number according to their specificity groups (mannose, galactose, N-acetylglucosamine, N- acetylgalactosamine, L-fucose, and N-acetylneuraminic acid) and to the monosaccharide that is the most effective inhibitor of the agglutination of erythrocytes or precipitation of carbohydrate- containing polymers by the lectin.[72].

The detailed specificities of a variety of lectins towards N-linked oligosaccharides have been elucidated and have been proposed as valuable tools in the fractionation and structural assessment of oligosaccharides and glycopeptides [73, 74, 75].

Of known lectins which have been purified and characterized, few bind sialic acid [76]. Wheat germ agglutinin (WGA) is the only plant lectin that binds to sialic acids. However, it also binds to oligosaccharides containing terminal N-acetylglucosamine and N-acetylgalactosamine[77].

Two lectins specific for sialic acids have been purified from animal sources, namely from the hemolymph of the horseshoe crabs of the class arachinda [78] and slug, Liman flavus [79].

Recently, two sialic acid-specific lectins have been purified from tritrichomonas mobilensis [80] and from the mushroom hericium exinaceum [81].

1.11.2 VERTEBRATE LECTINS

Vertebrate lectins can be subdivided on the basis of whether or not they are integrated into membranes.

a) Integral membrane lectins, that require detergents for their extraction (solubilization).

b) Soluble lectins.

This subdivision probably reflects a fundamental difference in the general functions of these classes[71].

The first group consists of lectins that differ in their sugar specificities (mannose, L-fucose, mannose-6-phosphate, N-acetylgalactosamine) and physiochemical properties[82].

Among the best-characterized lectins of this class are the galactose and mannose/N-acetylglucosamine-specific receptors of mammalian and avain hepatocytes, respectively, and the receptor for mannose-6-phosphate present in a variety of cells[83].

A galactose-specific lectin has been purified from extracts of metastatic B16 melanoma cells[84]. Membrane lectins seem to be present on the surfaces of various other cells, such as mouse and human lymphocytes, 3LL Lewis lung carcinoma, and mouse leukemia L1210[85], none of these lectins have been isolated.

The integral membrane lectins appear to have evolved to bind glycoconjugates to membranes, either at the cell surface or within vesicles. This results in the localization of the glycoconjugates at particular membrane sites or their transport to other cellular compartments[86].

In contrast, soluble lectins, being excluded from membranes, cannot directly function in this way. Instead they can move freely in aqueous compartments within and between cells, interacting with both soluble and membrane-bound glyacoconjugates..

Most of the soluble lectins isolated from vertebrate tissues bind β-galactosides. The best studied are a group of dimeric proteins found in many organisms, including the electric eel, calf, chicken, rat and man.

The concentration of these lectins in tissues is often ontogenically regulated, reaching a maximum at a given stage of development. All galactoside-specific lectins share not only similar subunit molecular weights and a dimeric structure, but also a requirement for a reducing agent to maintain their carbonhdrate binding activity. In the case of eel lectin, the reducing agent is required to prevent the oxidation of a tryptophan residue in the binding site[87].

Another group of vertebrate β-galactoside-binding lectins can be isolated as monomers. The one examined in most detail was purified from chicken intestine[88].

Other soluble vertebrate lectins are multimeric. The serum of the eel anguilla rostrata contains a lectin which has 12 subunits per molecule and is the first verterate lectin to have been characterized[89].

1.11.3 INVERTEBRATE LECTINS

Lectins are found in practically all of the approximately 30 phyla and the various classes and subclasses of invertebrates[90], mainly in the hemolymph and sexual organs, e.g albumin glands and eggs[91]. They appear also to present on the membranes of hemocytes[92] cells that function as primitive and rather unspecific immunological receptors.

Only a few of these lectins have been purified and characterized. Most of these lectins are specific for sialic acid.

1.11.4 BIOLOGICAL ROLES OF LECTINS

There is a considerable support, but little solid evidence, for the belief that lectins function primarily as recognition molecules. This function may be expressed differently in different organisms and also in different organs or tissues of the same organisms.

In plants, two proposed functions of lectins are currently attracting most attention:

a) As mediators of symbiosis between plants and microorganisms, for example, binding of nitrogen-fixing bacteria to legume roots. In this reaction bacteria of genus Rhizobium adhere to the surface of differentiated root cells and are then internalized into the root hair to form nitrogen fixing nodules [93].

b) Protection of plants against phytopathogenes:

The proposal that lectins may be involved in the defense of plants against fungal, bacterial, and viral pathogenes during germination and early growth of the seedlings is supported primarily by two lines of evidences:

1- The binding of lectins to various fungi and their ability to inhibit fungal growth and germination[94].

2- The presence of lectins at the potential sites of invasion by the infectious agents[95].

Membrane lectins are thought to mediate the binding of soluble extracellular and intracellular glycoproteins as well as of cells. The classical examples are:

• The binding of asialoglycoproteins by a galactose-specific lectin (receptor)on mammalian liver cells, and of asialioagalactoglycoproteins by mannose/N-acetylglucosamine –specific lectins (receptor) on avain hepatocytes. The complexes on the cell surfaces then undergo endocytosis and is transported to other cellular compartments; both interactions are probably key steps in the removal of these glycoproteins from the circulatory system[86] and/or replacement of sialic acid[96].

• Another example is pinocytosis of glycoproteins with terminal non-reducing mannose and/or N-acetylglucosamine residues by macrophages[97].

This uptake is mediated by a macrophage surface lectin specific for mannose and N-acetylgucosamine.

• The mannos-6-phosphate-specific lectin mediates the targeting of hydrolytic enzymes to the lysosomes [98].

• Galactose-specific lectins present on various human and murine tumors were suggested to influence the pathogenesis of cancer metastasis by promoting the formation of tumor cell aggregates (emboli) in the circulation and their adhesion to the endothelial layer of the capillaries[99].

The function of the soluble lectins has been more difficult to infer, partly because these lectins tend not to be as sharply localized as those in membranes.

A common function proposed for the soluble vertebrate lectins is to bind to the complementary glycoconjugates on and around the cells that release them [71] since both the lectins and glycoconjugates may be multivalent, large macromolecular complexes that shape extracellular environments may be formed [71].

This proposal is based on the finding that these proteins which are initially concentrated inside cells, are ultimately found extracellularly. For example, chicken lactose lectin I, which is concentrated intracellularly in developing muscle, becomes extracellular with maturation[100].

1.11.5 APPLICATIONS OF LECTINS

The bivalency or polyvalency of lectins and their specificity for particular sugar residues make them useful for many purposes[72]:

a) For blood typing.

b) As reagents for the study of simple and complex carbohydrates in solution and on cell surfaces.

c) For the identification and separation of cells.

d) For the selection of lectin-resistant mutants of animal cells with altered glycosylation patterns.

1.12 MULTIPLE MYELOMA (MYELOMATOSIS)

Multiple myeloma is a neoplastic monoclonal proliferation of bone marrow plasma cells. A single transformed plasma cell divides uncontrollably, consequently a very large number of cells of a single kind are produced, they form a clone because they are descended from the same cell with identical properties. Large amounts of immunoglobulin of a single kind are secreted by these tumors. Multiple myeloma is characterized by lytic bone lesions, plasma cell accumulation in the bone marrow; and the presence of monoclonal protein in serum and urine [101, 102].

1.12.1 IMMUNOPATHOLOGY

Normal plasma cells are derived from B-lymphocytes by transformation after exposure to antigenic stimuli; individual plasma cells manufactured immunoglobulins with only one type of light chain. The finding that in myeloma, all the malignant cells produce the same immunoglobulin indicates that the tumor is derived originally from one cell by cloning, the disease is therefore monoclonal [103].

The immunoglobulin is called a paraprotein and appears on electrophoretic strips as a clear-cut band. Each of the five normal types of immunoglobulin (IgG, IgM, IgA, IgE, and IgD) has light chains of either *lambda* or *kapa* variety [103].

In myeloma the paraprotein produced belongs to one of these immunoglobulin types and has one or other of the two light chains. In some patients only part of immunoglobulin molecule is produced by the tumor cells, most commonly the light chains. These appear in the urine as Bence-Jones proteinuria and if myeloma is associated only with light chains it is known as Bence-Jones myeloma [103].

The classification of myeloma by type of paraprotein and their relative frequency are given in table (1-2).

Patients with a myeloma which produces complete immunoglobulin molecules may also excrete increased amounts of light chain in their urine (Bence-Jones Proteinuria). In some this appears later as a new phenomenon and usually indicates an acceleration of the disease[103].

Table 1-2: Classification of multiple myeloma [103]

Type of paraprotein	Relative frequency %
IgG	55
IgA	21
Light chain	22
Other (D, E, nonsecretory)	2

1.12.2 EPIDEMIOLOGY

The disease is very uncommon under the age of 30, thereafter it becomes increasingly frequent, with a peak incidence between 60 and 70 years. Males are affected rather more frequently than females and Black people of Central African origin two to three times more often than Caucasians[103].

1.12.3 PATHOLOGY

In the majority of patients the bone marrow is heavily infiltrated with atypical plasma cells which are usually larger and paler staining than normal plasma cells and contain nucleoli, some cells may be multinucleated.

Progressive replacement of the marrow occurs with eventual reduction of the normal cell lines, inducing anaemia, leucopenia, and thrombocytopenia. Osteoclasts are stimulated and absorption of bone occurs, producing diffuse osteoporosis. Local tumor formation by the myeloma causes punched out translucencies in the bone radiograph. Rarely the disease may present as a solitary plasmacytoma either in bone or soft tissue[103].

Excessive production of the myeloma paraprotein is associated with progressive reduction in normal immunoglobulin levels and impairment of immune function[103].

1.12.4 CLINICAL FEATURES

There is a long preclinical phase, in some instances up to 25 years. The disorder may be discovered incidentally by laboratory tests during this phase and the patient may be observed for years before symptoms appear. The symptoms are[101]:

a) Bone pain (especially backache), pathological fractures.

b) Of anaemia: lethargy, weakness, dyspnoea, pallor, tachycardia, etc.

c) Repeated infections: these are related to deficient antibody production and, in advanced disease, to neutropenia.

d) Of renal failure and/or hypercalcaemia: polydipsia, polyuria, anorexia, vomiting, constipation and mental disturbance.

e) Abnormal bleeding tendency: myeloma proteins interferes with platelet function and coagulation factors; thrombocytopenia occurs in advanced disease.

f) Occasionally there is macroglossia, and diarrhoea due to amyloid disease.

g) Rarely there is a "Hyperviscosity syndrom" with purpura, haemorrhages, visual failure, CNS symptoms neuropathies, and heart failure. This results from polymerization of the abnormal immunoglobulin and is particularly likely when this is IgA, IgM or IgD.

1.12.5 DIAGNOSIS

A diagnosis of myeloma requires the detection of at least two of the following abnormalities[103].

a) Monoclonal immunoglobulin or light chains in blood or urine.

b) Infiltration of the marrow with malignant plasma cells.

c) Osteolytic bone lesions.

Other laboratory findings[101]

1- There is usually a normochromic normocytic or macrocytic anaemia. Neutropenia and throbocytopenia occur in advanced disease. Abnormal plasma cells appear in blood film in 15% of patients.

2- Very high erythrocyte sedimentation rate.

3- Serum calcium elevation occurs in 45% of patients. There is a normal serum alkaline phosphatase.

4- The blood urea is raised above 14 mmol/L and serum creatinine raised in 20% of cases. Proteinaceous deposits from heavy Bence-Jones proteinuria, hypercalcaemia, uric acid, amyloid and pyelonephritis may all contribute to renal failure.

5- A low serum albumin occurs with advanced disease

1.12.6 TREATMENT

Chemotherapy: Alkylating agents relieve pain, reduce plasma cell proliferation in the marrow and so reduce the serum paraprotein levels. As plasma cells are killed, normal bone marrow function improves. Melphalan or cyclophosphamide, with or without predinsolone, are the drugs of choice. Patients eventually become resistant to treatment, other drugs are then often tried, e.g. vincristine, adriamycin and bleomycin[101].

Radiotherapy: Radiotherapy is uniquely useful in the treatment of local problems such as severe bone pain, pathological fractures and tumorous lesions. Hemi-body irradiation in which the lower half of the body and then 6-8 weeks later the upper half is irradiated, may be employed for disseminated skeletal pain but is moderately toxic to bone marrow and lungs [103].

Prognosis: The median survival is two years with a 20% four years survival. The most serious prognostic feature is the blood urea concentration, if the blood urea is more than 14 mmol/L at presentation the median survival is only a few months. If the blood urea is less than 7 mmol/L at presentation the median survival is 33 months. Severe anaemia, a low serum albumin at presentation, and heavy Bence-jones proteinuria are also bad prognostic features[101].

1.12.7 BIOCHEMICAL ASPECTS IN MULTIPLE MYELOMA

1.12.7.1 SERUM INTERLEUKIN 2 (IL-2)

Serum IL-2 levels in 61 patients with multiple myeloma were analyzed. It has been shown that patients serum IL-2 levels were significantly higher than normal controls. Moreover, higher serum IL-2 levels were associated with a prolonged survival, in particular, 87% of the patients with IL-2≥10 U/mL are still alive at 5 years while only 13% of the remaining patients with IL-2<10 U/mL are alive, the data indicated that high serum IL-2 levels are the most useful predictor index of longer survival in MM patients. It has also been observed that all patients with IL-2≥10 U/mL had serum beta-2-microglobulin less than 6 µg/mL. Whereas in patients with IL-2<10 U/mL. SB2M ranged from 1.3 to 15 µg/mL. Using these two parameters, three groups of patients were classified with different survival duration. Group A had serum IL-2≥10 U/mL, SB2M<6 µg/ml in which all patients are alive. Group B, had IL-2<10 U/mL, SB2M<6 µg/mL, 42% of patients are alive. Group C, had serum IL-2<10 U/mL, SB2M≥6 µg/mL in which the survival curve drops to zero at 2.5 years. These data could reflect the existence of an active T cell control on B cell neoplasia[104].

1.12.7.2 THE NEURAL CELL ADHESION MOLECULE (NCAM)

The neural cell adhesion molecule (NCAM) is a membrane glycoprotein and belongs to the immunoglobulin superfamily. It is expressed on neural cells as well as on various neuroendocrine tumors and can be detected in sera of patients with small cell lung cancer. Its role is attributed to tumor invasion and formation of metastasis. Malignant plasma cells exhibits surface expression of

NCAM whereas normal plasma cells do not express NCAM. The expression dose not seem to correlate with clinical course, however myeloma cell lines tend to loose NCAM surface expression. An isoform of NCAM which is rich in polysialic acids has been shown to be elevated in sera of patients with multiple myeloma using a chemiluminescence immunoassay. Patients with progressive myeloma tend to have high serum NCAM levels above the normal range of 20 U/mL. Analysis of 125 myeloma patients suggest that serum NCAM is valuable parameter for tumor progression rather than tumor mass. Increase in serum NCAM may be associated with loss of adhesive function[105].

1.12.7.3 ACUTE PHASE PROTEINS

Serum IL-6 levels have been shown to be correlated with disease severity and prognosis in patients with plasma cell malignancies. Among its pleiortropic actions, IL-6 is also the major regulator of the acute phase proteins in humans. The possible effect on survival of the major serum acute phase proteins [c-reactive protein (CRP), alpha-1-antitrypsin (AAT) haptoglobin, acid alpha-1-glycoprotein and alpha-2-macroglobulin] was assessed on a population of 103 untreated myeloma patients. The analysis showed that among the acute phase proteins only AAT and CRP were significantly correlated with survival. Multivariate analysis showed that beta-2-microglobulin, calcium, creatinine, age, and AAT correlated significantly with survival. Combining B2M and AAT allowed stratification of myeloma patients, those with low levels of B2M and AAT presented on excellent prognosis, while those presenting higher values of the two parameters presented a median survival of 2.5 years[106].

1.12.7.4 SIALYL SALIVARY-TYPE AMYLASE

In 1988, it has been reported that there is an ectopic production of salivary-type amylase by an IgA-type multiple myeloma. Subsequently, seven cases of amylase-producing multiple myeloma have been reported. A neuraminidase sensitive, so-called sialyl salivary-type amylase were detected in the sera of two patients with IgA-type myeloma. The amylase showed an abnormal anodic migration in isoamylase electrophoretic analysis. The abnormal isoamylase could be separated from patient's sera by using size-exclusion HPLC. After the study of treatment with neuraminidase, the serum abnormal isoamylases were showed to be sialic acid residues-containing amylase. The two patients had no detectable malignancies except multiple myeloma, these findings strongly suggest that the sialic acid-containing salivary type amylases were produced ectopically from myeloma cells[107]. In a recent study, sialyl salivary-type amylasemia has been shown to be associated with IgD-type multiple myeloma[108].

1.12.7.5 SIALYLTRANSFERASE

A significant elevation of peripheral blood mononuclear cell sialyltransferase activity (STA) was observed in 14 patients with multiple myeloma, and compared to seven patients with monoclonal gamopathy of undetermined significance (MGUS) and to 10 controls. Serum sialyltransferase was significantly higher in MM patients as compared to controls. It was also higher than in (MGUS) patients, but the difference here was not statistically significant. (STA) was also determined in mononuclear bone marrow cells of 5 patients with MM, and found to be 19 times higher than that of bone marrow mononuclear cells from 6 patients with non-malignant disorder. No significant differences were observed in peripheral blood mononuclear cell sialic acid levels between MM and MGUS and controls[109].

1.12.7.6 SIALIC ACID

Sialic acid level has been determined recently in the sera of thirty-four patients with multiple myeloma. Nineteen of the myeloma patients had an IgG myeloma, eight IgM, six IgA and one had a lambda light chain myeloma. Serum total sialic acid was assayed by the enzymatic procedure. There was a highly significant difference between the serum sialic acid in the myeloma patients compared to the control group; 103.5 ± 32.8 mg/dL *versus* 78.9 ± 10.2 mg/dL respectively ($P<0.001$). There was no difference in the serum sialic acid between the three major groups of the myeloma patients (i.e. grouped by the immunoglobulin type). When the myeloma patients were taken as one group, regardless of the type of myeloma paraprotein present, there was no significant correlation between serum sialic acid and the amount of paraprotein, nor was there a significant correlation between serum sialic acid and serum total protein, albumin or globulin. There was a significant correlation between serum sialic acid and serum total protein and globulin concentration in the IgA myeloma patients, and significant correlation between serum paraprotein quantity and globulin concentration in the IgM patients. Conversely, there was no correlation between serum sialic acid and any of the other serum variables in the IgG myeloma patients[110].

AIM OF WORK

The aim of the work in this thesis includes the following:

1- Determination of TP,TSA/TP ratio and LBSA in sera of multiple myeloma, leukemia and breast cancer patients, and to investigate the possibility of using them as a tumor markers by comparison with their levels in normal individuals.

2- Determination of serum levels of mucoid proteins and protein-bound hexoses in multiple myeloma , leukemia and breast cancer patients and to compare these levels with those from normal individuals.

3- Measurement of SOD activity in sera of patients with multiple myeloma, leukemia and breast cancer and to compare these values with the normal values from healthy individuals.

4- Molecular characterization and determination of the optimum conditions for the binding of lectin to human erythrocyte surface glycoconjugates through the study of the effect of various factors.

5- Determiantion of the kinetic and thermodynamic parameters of lectin binding to human erythocyte surface glycoconjugates by using multiple myeloma serum sample.

Experimental

2.1 CHEMICALS, INSTRUMENTS, AND PATIENTS

2.1.1 CHEMICALS

All common laboratory chemicals or reagents were of analr grade or the equivalent, and were used without further purification. They were obtained from the following companies:-

a- BDH Company

Na,k-tartrate, Folin-cio calteau, Chloroform. Butylacetate, HCl, Perchloric acid, H_2SO_4, Benzoic acid, Mannose, Fructose, NBT, Riboflavin, EDTA, NaCl, NaOH, $CaCl_2$ KCN and Neuraminidase.

b- Fluka Company:

Ethanol, Methanol, Galactose, Orcinol, Tris (hydroxy methylamino methan), and Bovine serum albumin (BSA).

c- May and Baker Company:

Resorcinol, Na_2HPO_4, and Urea.

d- Riedel-De Haenag Company:

Phosphotungstic acid, $CaCO_3$, $CuSO_4$. $5H_2O$, KH_2PO_4, and Xylose.

e- Merk Company: Polyethylene glycol.

f- Sigma Company. Sialic acid and D-Glucuronic acid.

2.1.2 INSTRUMENTS

a- LKB spectrophotometer ultraspec type 4050.

b- Pye-Unicam pH meter.

c- MSE Centrifuge.

d- Memert water bath.

e- SM-Shaker.

2.1.3 PATIENTS

Seventy two patients with cancer diseases and thirty normal donors were included in this study. They were classified into four groups: group I ,22 patients with known multiple myeloma. The diagnosis was based on a number of features including the production of a paraprotein, the presence of malignant plasma cells within the bone marrow and evidence of bone destruction on X-ray examination.

group II, 25 patients with leukemia. The diagnosis was based on the clinical features and blood film. group III, 25 patients with breast cancer the diagnosis was based on clinical feature and histologically by tissue biopsy. group IV, 30 normal healthy donors.

All patients were admitted for diagnosis and treatment to Saddam Medical City Hospital under the supervision of specialists Dr. Hassan Ibraheem, Dr. Ali Mohammed Jawad, and Dr. Shawky Yossef Fawzi. The patients were newly diagnosed and not underwent any type of therapy.

The host information of the patients and healthy persons is summarized in table (2-1).

Table (2-1): The host information of all patients and healthy subjects studied.

Group	Case	NO.of cases	Female	Male	Age (year)
I	Multiple myeloma	22	14	8	60±5.3
II	Leukemia	25	13	12	32±6.4
III	Breast cancer	25	25	-	38±4.4
IV	Normal control	30	9	21	28±3.8

serum specimens

Blood samples were collected from the patients and healthy donors by venipuncture. The whole blood was left for 20 minutes at room temperature. After coagulation, the serum was separated by centrifugation at 3000 r.p.m. for 10 minutes, serum specimens were frozen at $-20°C$ until assayed.

2.2 DETERMINATION OF BIOCHEMICAL CONSTITUENTS IN SERA OF PATIENTS WITH MULTIPLE MYELOMA, LEUKEMIA, AND BREAST CANCER.

2.2.1 DETERMINATION OF TOTAL PROTEIN (TP) IN SERA OF PATIENTS WITH MULTIPLE MYELOMA, LEUKEMIA, AND BRESAT CANCER.

Protein content of the serum was determined by the Lowry method, using bovine serum albumin (BSA) as the standard protein[111].

Reagents

1- Reagent A: Alkaline sodium carbonate solution 2gm. of sodium carbonate (Na_2CO_3) was dissolved in 100mL of 0.1N sodium hydroxide (NaOH).

2- Reagent B: Copper sulphate-Na,K-tartrate solution. 0.5gm of copper sulphate ($CuSO_4.5H_2O$) was dissolved in 100mL distilled water. From this solution, 10mL was taken to be added to 0.1gm of Na,K-tartrate. This reagent was prepared freshly in the day of use.

3- Reagent C: Alkaline copper solution: This reagent is prepared on the day of use by mixing 50mL of reagent A and 1mL reagent B.

4- Reagent D: Folin-cio calteau reagent: This reagent is a solution of sodium tungstate and sodium molybdate in phosphoric acid and hydrochloric acid. The reagent was prepared by dilution of the commercial reagent with an equal amount of distilled water on the day of use.

5- Standard protein: The standard protein (BSA), was prepared as follows:

   a- A stock solution of 1000 µg/mL was prepared by dissolving 100mg BSA in 100 mL distilled water.

   b- From the stock solution, the following concentrations were prepared by serial dilutions with distilled water: 25,50,100,150,200 µg/mL.

Procedure

1- 2.5mL of reagent C was added to 0.5 mL of standard protein or diluted serum sample and left for 10 minutes after mixing them thoroughly.

2- 0.25mL of reagent D was added and mixed immediately and rapidly. This mixture was left for 30 minutes, then the absorbance was read at 600nm.

Calculations

The standard curve was obtained by plotting the absorbance against the corresponding concentrations of standard protein, and was used to determine the unknown protein concentration of the serum sample.

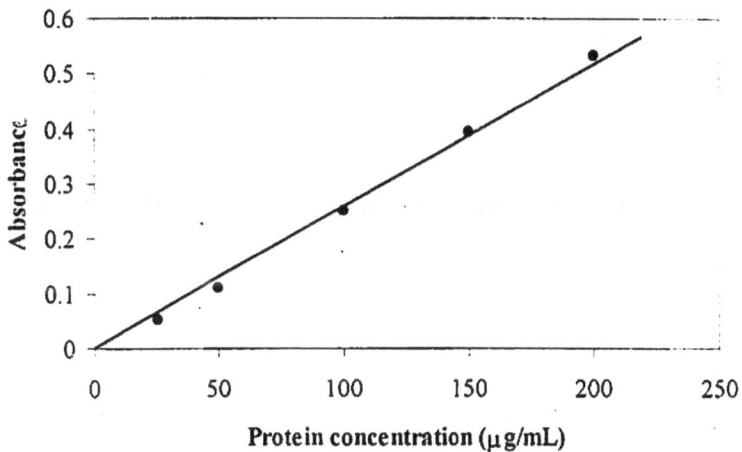

Figure (2-1): The standard curve of protein determination by the Lowry method.

2.2.2 DETERMINATION OF TOTAL SIALIC ACID (TSA) LEVELS IN SERA OF PATIENTS WITH MULTIPLE MYELOMA, LEUKEMIA, AND BREAST CANCER.

Reagents

1- Resorcinol stock solution (2%w/v): 1gm of resorcinol was dissolved in 50mL distilled water. The solution was prepared in the day of use.

2- Copper sulphate solution (0.1M): 4gm of $CuSO_4.5H_2O$ was dissolved in 250mL distilled water.

3- Resorcinol reagent: This reagent was prepared in the day of use by mixing 5mL of resorcinol stock solution, 4.875mL distilled water, and 0.125mL (0.1M) $CuSO_4.5H_2O$ solution, the final volume was made up to 50mL with HCl (concentrated).

4- Butylacetate /Methanol reagent (85:15v/v): 85mL of butylacetate was mixed with 15mL methanol. This solution was stored cold.

5- Standard sialic acid: The standard sialic acid solutions with different concentrations (5,10,15,20,25,30,35,40,45μg/mL) were prepared by serial dilutions from a stock standard solution of sialic acid.

Procedure

1- 980 μL of distilled water was added to 20μL of standard sialic acid solutions or serum sample, the assay tubes were vortexed and placed on ice.

2- To each assay tube, 1mL of resorcinol reagent was added, then the tubes were placed in a boiling water for exactly 15 minutes, then for 10 minutes on ice.

3- 2mL of butylacetate/ methanol reagent was added, the assay mixtures were vortexed and centrifuged for 10 minutes at 3000 r.p.m. The extracted chromophore was read at 580nm. against distilled water.

Calculations

The standard curve was obtained by plotting the absorbance at 580nm. against the corresponding concentrations of standard sialic acid solutions, and was used to determine the TSA levels in the serum samples.

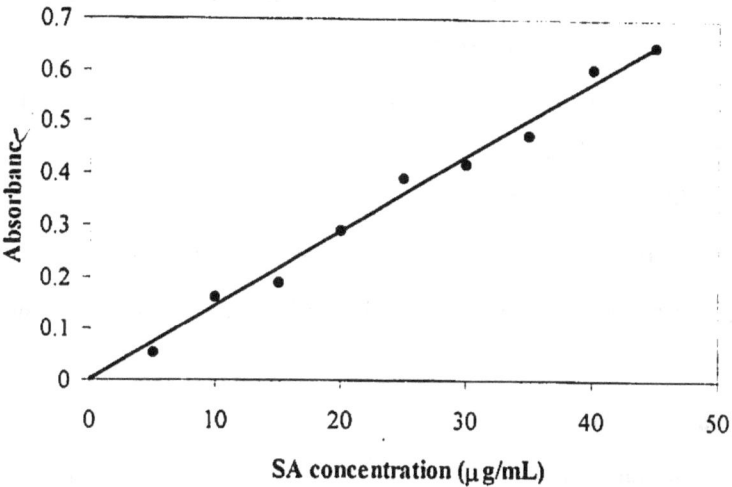

Figure (2-2): The standard curve for the determination of sialic acid concentrations in human serum.

2.2.3 DETERMINATION OF LIPID-BOUND SIALIC ACID (LBSA) LEVELS IN SERA OF MULTIPLE MYELOMA, LEUKEMIA, AND BREAST CANCER PATIENTS.

Reagents

1- Chloroform/Methanol: This solution was prepared by mixing two volumes of chloroform with one volume of methanol, it was stored at 4-5°C.

2- Phosphotungstic acid solution: 1gm of phosphotungstic acid was dissolved in 1mL of distilled water, then heated to produce clear solution, the solution was prepared freshly in the day of use.

3- Resorcinol reagent and butylacetate/methanol reagent were prepared as described in section (2.2.2).

Procedure

LBSA was assayed in serum samples by using the method described by Katopodis et.al.[112]. LBSA assay consists of the following steps:

1- 150μL of distilled water was added to 50μL of serum sample, the tubes were vortexed for five seconds and placed on ice.

2- 3mL of cold (4°C) chloroform/methanol solution was added, after mixing them, 0.5mL of cold distilled water was added, the tubes were centrifuged for five minutes at 3000 r.p.m. at room temperature.

3- One mL of the resulting upper layer was transfered to another tube, and 50μL of phosphotungstic acid solution were added, mixed and allowed to stand at room temperature for 5 minutes.

4- The assay tubes were centrifuged for 5 minutes at 3000 r.p.m. the supernatant fluid was removed and the remaining precipitate was dissolved in one mL of distilled water.Then the steps 2 and 3 mentioned in section (2.2.2) were followed.

Calculations

LBSA levels in serum samples were determined using the standard sialic acid curve (Figure.2-2).

2.2.4 DETERMINATION OF GLYCOPROTEIN LEVELS IN THE SERUM SAMPLES OF MULTIPLE MYELOMA, LEUKEMIA, AND BREAST CANCER PATIENTS.

2.2.4.1 DETERMINATION OF MUCOID PROTEINS (MP) IN SERA OF MULTIPLE MYELOMA, LEUKEMIA, AND BREAST CANCER PATIENTS.

Reagents

1- Perchloric acid solution (1.8M): 16.6mL of 72% perchloric acid were diluted to 100mL with distilled water.

2- Phosphotungstic acid solution: 5gm of phosphotungstic acid were dissolved in 100mL of (2N) HCl.

3- Sodium chloride solution 0.85%.

4- Sodium hydroxide solution (0.1N): 0.4 gm of NaOH was dissolved in 100mL of distilled water.

5- H_2SO_4 60%v/v: 60mL of H_2SO_4 were mixed with 40mL of distilled water.

6- Orcinol reagent: 2gm of orcinol were dissolved in 100mL of 30%v/v H_2SO_4.

7- Stock standard sugars: 100mg of galactose and 100mg of mannose were dissolved in 100mL of distilled water, then saturated with benzoic acid and stored cold.

8- Working standard: 1mL of stock standard sugars was mixed with 9mL of distilled water, this reagent was prepared freshly in the day of use.

Procedure

Mucoid proteins (MP) were estimated by using the method of Weimer and Mashin[113]. The method includes the following steps:

1- 100μL of serum sample were added to 900μL of NaCl solution, mixed, then 500μL of (1.8M) perchloric acid were added, after mixing by inversion, the assay tubes were allowed to stand at room temperature for 10 minutes, then centrifuged for 15 minutes at 3500 r.p.m. to obtain clear supernatant.

2- To one milliliter of the supernatant, 200μL of phosphotungstic acid solution were added, after mixing, the tubes were allowed to stand for 10 minutes.

3- The tubes were centrifuged, after removing of the supernatant, one milliliter of 95% ethanol stire was added, centrifuged and the supernatant was removed.

4- The resulting precipitate was dissolved in 100μL of (0.1N) NaOH, this was considered as the unknown.

5- The standard was set up by using 100µL of working standard, and the blank by using 100µL of distilled water.

6- 250µL of orcinol reagent and 1.5mL of 60% H_2SO_4 were added to unknown, standard and blank.

7- All tubes were placed in a water bath of ($80\pm0.5°C$) for 20 miuntes, cooled and read against distilled water at 520nm.

Calculations

The calculations were done as follows:-

$$MP \ (mg \ / 100 \ mL \ ) = \frac{A_x - A_b}{A_s - A_b} \times 0.1 \times \frac{100}{0.333}$$

$$= \frac{A_x - A_b}{A_s - A_b} \times 30$$

Where:

A_x: the absorbance of unknown solution at 520nm.

A_s: the absorbance of standard solution at 520nm.

A_b: the absorbance of blank solution at 520nm.

2.2.4.2 DETERMINATION OF PROTEIN-BOUND HEXOSES IN SERA OF PATIENTS WITH MULTIPLE MYELOMA, LEUKEMIA, AND BREAST CANCER.

Reagents

All reagents were prepared as described in section (2.2.4.1)

Procedure

1- 20µL of serum were added to one milliliter of 95%v/v ethanol and mixed.

2- The mixture was centrifuged and the supernatant was discarded, the remaining precipitate was washed with 1mL of 95% ethanol, centrifuged, after removing the supernatant, the steps 4-7 described in section (2.2.4.1) were performed[113].

Calculations

The calculations were done as follows:

$$Potein - bound\ hexoses\ (mg/100ml) = \frac{A_x - A_b}{A_s - A_b} \times 0.1 \times \frac{10}{0.1}$$

$$= \frac{A_x - A_b}{A_x - A_b} \times 100$$

Where:

A_x: the absorbance of unknown solution at 520nm.

A_s: the absorbance of standard solution at 520nm.

A_b: the absorbance of blank solution at 520nm.

2.3 DETERMINATION OF SUPEROXIDE DISMUTASE (SOD) ACTIVITY IN SERA OF PATIENTS WITH MULTIPLE MYELOMA, LEUKEMIA, AND BREAST CANCER.

Superoxide dismutase activity was measured by the method of Winterbourn et al[114]. with our modification for serum.

Reagents

1- Phosphate buffer (0.067M) pH=7.8 This buffer was prepared by mixing the appropriate amounts of 0.067M KH_2PO_4 stock solution and 0.067M Na_2HPO_4. $12H_2O$ stock solution, the final volume was made up to 100mL with distilled water.

2- EDTA/KCN solution: This solution was prepared by dissolving 1.5mg of KCN in 100mL of (0.1M) EDTA solution.

3- Riboflavin solution (0.12mM): 4.5mg of riboflavin were dissolved in 100mL of distilled water, stored cold in a dark bottle.

4- NBT solution (1.5mM): 12.3mg of NBT were dissolved in 10mL of distilled water, stored cold.

Procedure

1- 200µL of EDTA/KCN solution were added to 100µL of serum sample, then 100µL of NBT solution were added.

2- The assay tubes were brought to (20-22°C), after that 50µL of riboflavin solution were added to each tube. The final assay volume of 3mL was made up with phosphate buffer 0.067M pH=7.8 .

3- Subsequent exposure to bright lighting was controlled by placing the assay tubes in a white-light box where they received uniform illumination for 20 minutes with 18W fluorscent tube, then the absorbance was read at 560nm. against distilled water.

4- To determine the control value, the absorbance for another set of tubes containing the same mixture was read at 560nm. against distilled water immediately after the addition of riboflavin. (riboflavin was added after the addition of buffer).

5- To determine SOD unit, six tubes containing 10,20,40,60,80, and 500µL of normal serum sample, and another tube containing no serum were treated as described in the steps 1,2, and 3.

Calculations

1- Percentage inhibition was calculated from each absorbance in the presence and absence of the enzyme:

Inhibition % = $(A_E - A_{NE}) \times 100$

where:

A_E: The absorbance at 560nm of the tubes containing different amounts of the enzyme.

A_{NE}: The absorbance at 560nm in the absence of the enzyme.

2- The percentages of inhibition were plotted against the corresponding amounts of serum (Figure(2-3)).

3- SOD unit was calculated from figure(2-3) according to the following: the amount of serum (VμL) which gives half the maximum inhibition of NBT reduction (1unit=10.1μL).

4- To calculate the SOD activity in sera of patients, the differences between absorbances before and after the light irradiation were multiplied by the SOD unit.

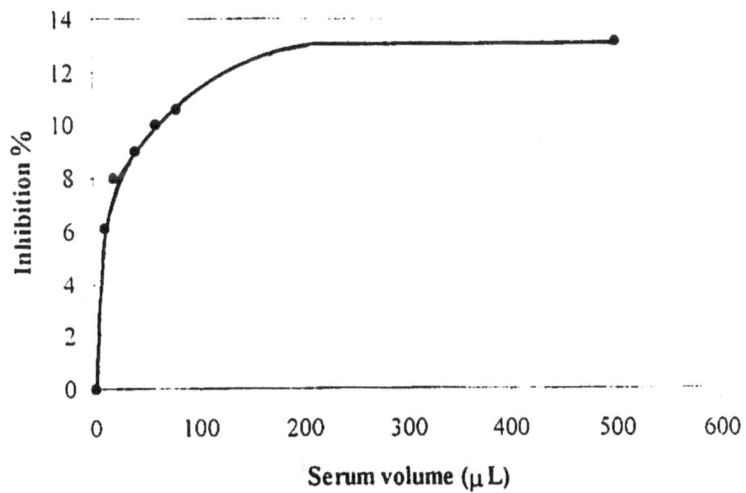

Figure (2-3): The standard curve for the determination of SOD unit.

2.4 BINDING STUDIES OF THE LECTIN TO ERYTHROCYTE SURFACE GLYCOONJUGATES IN MULTIPLE MYELOMA AND LEUKEMIA SERUM SAMPLES.

2.4.1 PRELIMINARY TEST FOR THE LECTIN BINDING TO ERYTHROCYTE SURFACE GLYCOCONJUGATES.

2.4.1.1 DETERMINATION OF TOTAL LECTIN BINDING TO ERYTHROCYTE SURFACE GLYCOCOJUGATES.

Reagents

1- Standard erythrocyte suspension: Fresh human erythrocytes (type A) were washed (3-4)times with an appropriate amounts of normal saline (0.9% NaCl solution), then the cells were diluted with normal saline to give an absorbance value of about 2 at 620 nm[115]. The suspension was prepared in the day of the assay.

2- Tris - HCl buffer:

(a) 0.2 M stock solution of tris (hydroxymethyl aminomethan): 2.4228 gm in 1000mL distilled water.

(b) 0.2 M HCl stock solution.

The assay tris-HCl buffer pH=8 containing 0.15M NaCl and 20mM $CaCl_2$ was prepared by dissolving 2.193 gm and 0.555gm of NaCl and $CaCl_2$ respectively in 250 mL of the buffer.

Procedure

The total binding of cancerous lectin to erythrocyte surface glycoconjugates was preliminary tested by the hemagglutination assay. This assay is a modification of the method described by Liener[116][117]. The assay includes the following steps:

1- 0.5 mL of erythrocyte suspension was added to 50μL of serum (cancerous lectin), then 450μL of Tris-HCl buffer pH=8 were added to give a final volume of 1mL. The assay mixture was incubated for 30 minutes at room temperature.

2- The assay mixture was centrifuged for 5 minutes at 2000 r.p.m., then the supernatant was removed.

3- The remaining red blood cells were resuspended in 500μL of tris-HCl buffer pH=8. The reacted (bound) cells were allowed to sediment for 30 minutes, then the absorbance for the upper layer (free lectin and cells) was read at 620nm.

Calculations

Total binding represents the amount of lectin which binds the erythrocytes surface glycoconjugates causing hemagglutination.

$$TB \% = \frac{A - A^*}{A} \times 100$$

where:

$TB\%$: The percent of total binding of lectin to erythrocyte surface glycoconjugates.

A : The absorbance of standard erythrocyte suspension at 620 nm.

A^*: The absorbance of free (unbound) erythrocytes at 620nm.

2.4.1.2 DETERMINATION OF NON-SPECIFIC BINDING OF LECTIN TO ERYTHROCYTE SURFACE GLYCOCONJUGATES.

Reagents

The buffer solution was prepared as described in section (2.4.1.1) and the erythrocyte suspension used in this assay was prepared as follows:

a- The erythrocytes were washed (3-4) times with normal saline (0.9 % NaCl), then two times with assay buffer (tris-HCl buffer pH8). The suspension was prepared.

b- 100μL of neuraminidase (500 unit/mL) were added to 10mL of the suspension. The mixture was shaken for four hours at 25°C.

c- After centrifugation, the supernatant was removed and the remaining red blood cells were diluted with an appropriate amount of normal saline to give an absorbance value of about 2 at 620nm.

Procedure

The same steps described in section (2.4.1.1) were followed to determine the percent of the non- specific binding.

Calculations

The percent of non-specific binding was calculated using the following equation:

$$NSB \ \% \ = \ \frac{A^{\bullet} - A^{*}}{A^{\bullet}} \times 100$$

where:

$NSB\%$: The percent of non-specific binding.

$A^{\bullet}$: The absorbance of neuraminidase-treated erythrocyte suspension at 620nm.

A^{*} : The absorbance of free (unbound) erythrocytes at 620nm.

2.4.1.3 DETERMINATION OF THE SPECIFIC BINDING OF LECTIN TO ERYTHROCYTE SURFACE GLYCOCONJUGATES.

The percent of specific binding of lectin to glycoconjugates was calculated by subtracting the percent of non- specific binding from the percent of the total binding.

$SB\% = TB\% - NSB\%$

where:

$SB\%$: The percent of specific binding of lectin to erythrocyte surface glycoconjugates

2.4.2 FACTORS EFFECTING ON LECTIN BINDING TO ERYTHROCYTE SURFACE GLYCOCONJUGATES IN MULTIPLE MYELOME AND LEUKEMIA.

2.4.2.1 THE EFFECT OF DIFFERENT LECTIN AMOUNTS ON ITS BINDING TO ERYTHROCYTE SURFACE GLYCOCONJUGATES.

Reagents

The standard erythrocyte suspension and tris-HCl buffer were prepared as described in section (2.4.1.1).

Procedure

0.5 mL of erythrocyte suspension was added to 100µL of increasing amounts (375, 750, 1125, 1500, 1875, 2250, 2625, 3000, 3375, 3750 and 4500 µg) of protein in a total volume of 1mL (completed with tris-HCl buffer pH=8). After incubation for 30 minutes at 25°C, the steps 2 and 3 described in section (2.4.1.1) were followed.

Calculations

1- The percent of total binding was calculated according to the formula mentioned in section (2.4.1.1).

2- The percent of specific binding (SB%) was calculated by using the equation mentioned in section (2.4.1.3).

3- The percent of specific binding was plotted against the corresponding protein amount included in each mixture.

2.4.2.2 THE EFFECT OF pH ON LECTIN BINDING TO ERYTHROCYTE SURFACE GLYCOCONJUGATES.

Reagents

Five buffer solutions(tris-HCl buffer) with different pH values (7, 7.5, 8, 8.5, 9) were prepared, each buffer contains 0.15M NaCl and 20mM $CaCl_2$, section (2.4.1.1).

Procedure

0.5mL of erythrocyte suspension was added to 50μL of serum (3750 μg protein), the volumes of the mixtures were made up to 1mL with tris-HCl buffer of different pH values (7, 7.5, 8, 8.5, 9), the assay tubes were incubated for 30 minutes at 25°C. After incubation the steps 2 and 3 of section (2.4.1.1) were repeated.

Calculations

1- The percent of total binding (TB%) was determined by using the same formula mentioned in section (2.4.1.1) for each pH value.

2- The percent of specific binding (SB%) was calculated using the formula mentioned in section (2.4.1.3).

3- The percentages of specific binding (SB%) were plotted against the corresponding pH values.

2.4.2.3 THE EFFECT OF TEMPERATURE ON BINDING OF LECTIN TO ERYTHROCYTE SURFACE GLYCOCONJUGATES

Reagents

The erythrocyte suspension and buffer solution were prepared as described in section (2.4.1.1).

Procedure

0.5mL of erythrocyte suspension was added to 50µL of serum (3750 µg protein). The final assay volumes were made up to 1mL with tris-HCl buffer (pH=8.5) which contains 0.15M NaCl and 20 mM $CaCl_2$. The tubes were incubated for 30 minutes at different temperatures (5, 10, 15, 20, 25, 30, 37 °C). After incubation, the experiment was performed according to the steps 2 and 3 described in section (2.4.1.1).

Calculations

1- The percent of total binding (TB%) was estimated using the mathematical formula mentioned in section (2.4.1.1) for each temperature.

2- The percent of specific binding (SB%) was determined according to the equations described in section (2.4.1.3) for each temperature.

3- The percentages of specific binding (SB%) were plotted against the corresponding temperatures.

2.4.2.4 THE EFFECT OF INCUBATION TIME ON LECTIN BINDING TO ERYTHROCYTE SURFACE GLYCOCONJUCATES

Reagents

The erythrocyte suspension and tris-HCl buffer were prepared as described in section (2.4.1.1).

Procedure

0.5 mL of erythrocyte suspension was added to 50 µL of serum (3750 µg protein), the final volumes were made up to 1mL with tris-HCl buffer (pH 8.5) which contains 0.15M NaCl and 20 mM $CaCl_2$. The assay tubes were incubated at 37 °C for (15, 30, 60, 90, 120 and 150 minutes), then the experiment was performed according to the steps 2 and 3 described in section (2.4.1.1).

Calculations

1- The percent of total binding (TB%) was calculated according to the formula mentioned in section (2.4.1.1) for each time.

2- The percent of specific binding (SB%) was calculated according to section (2.4.1.3) for each time.

3- The percentages of specific binding (SB%) were plotted against the corresponding times.

2.4.2.5 THE EFFECT OF EXOGENOUS Ca^{-2} IONS CONCENTRATION ON THE BINDING OF LECTIN TO ERYTHROCYTE SURFACE GLYCOCONJUGATES

Reagents

Five solutions of the assay buffer with different concentrations of Ca^{-2} ions (5, 10, 15, 20, 30 mM) were prepared by dissolving (1.38, 2.77, 4.16, 5.55, 8.32×10^{-2} gm) respectively, in 25 mL of tris-HCl buffer pH=8.5.

Procedure

0.5 ml erythrocyte suspension was added to 50 μL of serum (3750 μg protein), the final assay volumes were made up to 1mL with tris-HCl buffer pH=8.5 which contains different concentrations of Ca^{+2} ions (5, 10, 15, 20, 30 mM). The assay mixtures were incubated for 120 minutes at 37 °C. After incubation the experiment was performed according to steps 2 and 3 in section (2.4.1.1).

Calculations

1- The percent of total binding (TB%) was determined using the formula mentioned in section (2.4.1.1) for each tube.

2- The percent of specific binding (SB%) was calculated according to section (2.4.1.3).

3- The percentages of specific binding (SB%) were plotted against the corresponding Ca^{+2} concentrations.

2.4.2.6 THE EFFECT OF IONIC STRENGTH AND DIFFERENT SALTS ON BINDING OF LECTIN TO ERYTHROCYTE SURFACE GLYCOCONJUGATES

a- The effect of monovalent salt on lectin binding

Reagents

(7.31, 14.6, 21.9, 29.2, 43.8×10^{-2} gm) of NaCl were each dissolved in 25mL of tris-HCl buffer pH=8.5 to prepare five buffer solutions contain the following NaCl concentrations: (50, 100, 150, 200 and 300 mM) respectively.

Procedure

0.5 mL of erythrocyte suspension was added to 50 µL of serum (3750 µL protein), the final volumes were completed to 1mL with tris-HCl buffer pH=8.5 which contains different NaCl concentrations (50, 100, 150, 200 and 300 mM). After incubation for 120 minutes at 37°C the experiment was performed according to the steps 2 and 3 described in section (2.4.1.1).

Calculations

The percent of total binding (TB%) and the percent of specific binding (SB%) were calculated according to sections (2.4.1.1) and (2.4.1.3) respectively. The percentages of specific binding were plotted against the corresponding NaCl concentrations.

b- The effect of divalent salt on lectin binding.

Reagents

(2.47, 4.94, 7.42 and 9.89×10^{-2} gm) of MgCl$_2$ were each dissolved in 25mL of tris-HCl buffer pH=8.5 to prepare buffer solutions contain (5,10,15 and 20×10^{-3} M) of MgCl$_2$.

Procedure

0.5 mL of erythrocyte suspension was added to 50 μL of serum (3750 μg protein), the final volumes of assay mixtures were completed with the assay buffer which contains the following $MgCl_2$ concentrations (5, 10, 15 and 20×10^{-3} M). The tubes were incubated for 120 minutes at 37 °C then the steps 2 and 3 described in section (2.4.1.1) were repeated.

Calculations

The percent of total binding (TB%) and the percent of specific binding (SB%) were calculated for each assay tube according to sections (2.4.1.1) and (2.4.1.3) respectively. The percentages of specific binding (SB%) was plotted against the corresponding $MgCl_2$ concentrations.

2.4.2.7 THE EFFECT OF DIFFERENET DENATURATING AGENTS ON LECTIN BINDING TO ERYTHROCYTE SURFACE GLYCOCONJUCATES

Reagents

Urea, polyethyleneglycol, NaOH and HCl were used as denaturating agents in this study. Solutions of urea were prepared in different concentrations (3, 4, 5 and 6 M) by dissolving (4.5, 6, 7.5 and 9 gm) of urea in 25 mL of tris-HCl buffer pH=8.5. Another set of solutions with different concentrations of polyethylene glycol (0.5%, 1%, 2% and 4%) was prepared by dissolving (0.125, 0.25, 0.5 and 1 gm) of PEG in 25mL tris-HCl buffer pH 8.5.

Sodium hydroxide solution (0.15M) was prepared by dissolving 0.15g in 25mL tris-HCl buffer pH 8.5. Hydrochloric acid solution (0.15 M) was prepared by the addition of 98.69mL tris-HCl-buffer pH 8.5 to 1.31 mL of HCl.

Procedure

0.5 mL of erythrocyte suspension was added to 50 µL of serum (3750 µg protein), the final assay volume of 1mL was made up with tris-HCl buffer pH 8.5, which contains different concentrations of the following denaturating agents (Urea, PEG, NaOH and HCl). The tubes were incubated for 120 minutes at 37 °C. After incubation they were treated as previously described in step 2 and 3 of section (2.4.1.1).

Calculation

1- The values of the percent of total binding (TB%) were determined for each assay. the formula in section (2.4.1.1) was used.

2- The corresponding values of the percent of specific binding (SB%) were determined according to section (2.4.1.3).

3- The percentages of specific binding (SB%) was plotted against the corresponding concentrations of urea and PEG. Other results which obtained using NaOH and HCl were tabled.

2.4.3. INHIBITION STUDIES OF LECTIN BINDING TO ERYTHROCYTE SURFACE GLYCOCONJUGATES USING MULTIPLE MYELOMA AND LEUKEMIA SERUM SAMPLES

The inhibition studies were performed by using different concentrations of sialic acid, glucuronic acid, furctose, mannose, and xylose, as inhibitors for the binding of lectin to glycoconjugates of erythrocyte surface. All experiments below were carried out at the optimum conditions of lectin amount 50µL (3750 µg protein), temperature 37°C, pH 8.5, incubation time 120 minutes, and Ca^{+2} concentration (15mM).

2.4.3.1 INHIBITION BY SIALIC-ACID

Reagents

A stock solution of 10mM of sialic acid was prepared. from this solution, the following concentrations (1, 1.5, 2, 2.5 and 3 mM) were prepared by serial dilutions with tris-HCl buffer pH 8.5.

Procedure

0.5mL of erythrocyte suspension was added to 50µL of serum (3750 µg protein). The assay volume was made up to 1mL with sialic acid solutions containing (1, 1.5, 2, 2.5 and 3 mM) of sialic acid. After incubation for 120 minutes at 37°C, the assay tubes were treated according to step 2 and 3 in section 2.4.1.1. The same steps were carried out for another set of tubes in the absence of the inhibitor.

Calculations

1- The percent of total binding (TB%) was calculated in the presence and absence of the inhibitor for each tube, the formula of section (2.4.1.1) was used.

2- The percent of specific binding (SB%) was calculated for each tube in the presence and absence of the inhibitor by using the equation in section (2.4.1.3).

3- The percent of inhibition (Inh%) was calculated as the difference between the percent of specific binding in the absence of inhibitor and that obtained in the presence of it.

$Inh.\% = SB\%_{(NI)} - SB\%_{(I)}$

Where:

Inh.% : The percent of inhibition by sialic acid.

$SB\%_{(NI)}$: The percent of specific binding in the absence of inhibitor.

$SB\%_{(I)}$: The percent of specific binding in the presence of inhibitor (sialic acid).

4- The percentages of inhibition were plotted against the correspond sialic acid concentrations.

2.4.3.2 INHIBITION BY GLUCURONIC ACID

Reagents

A stock solution of 50 mM of glucuronic acid was prepared, from this solution the following concentrations (1, 10 15, and 20 mM) were prepared by serial dilutions with tris-HCl-buffer pH 8.5.

Procedure

The same steps previously described in section (2.4.3.1) were followed except the glucuronic acid solution were used in this assay.

Calculations

1- The values of (TB%) were calculated in the presence and absence of the inhibitor for each tube, the equation of section (2.4.1.1) was used.

2- The values of (SB%) were calculated for each tube according to section (2.4.1.3).

3- The percent of inhibition (Inh.%) was calculated according to the following formula.

$Inh.\% = SB\%_{(NI)} - SB\%_{(I)}$

Where:

$Inh.\%$: The percent of inhibition by glucuronic acid.

$SB\%_{(NI)}$: The percent of specific binding in the absence of inhibitor.

$SB\%_{(I)}$: The percent of specific binding in the presence of inhibitor (glucuronic acid).

4- The Inh% values were plotted against the corresponding glucuronic acid concentrations.

2.4.3.3 INHIBITION BY FRUCTOSE, MANNOSE, AND XYLOSE

Reagents

(0.054, 0.054, 0.045 gm) of fructose, mannose, and xylose respectively were dissolved in 10mL of tris-HCl buffer pH 8.5 to prepare three solutions contain 30 mM of these sugars.

Procedure

0.5mL of erythrocyte suspension was added to 50µL of serum (3750 µg protein), the final volume of the first tube was made up with tris-HCl buffer pH 8.5 which contain 30mM of fructose, the second tube with buffer solution containing 30mM of mannose and the third one with buffer containing 30mM of xylose. The assay tubes were incubated for 120 minutes at 37 °C, then the steps 2 and 3 described in section (2.4.1.1) were followed.

Calculations

The calculations were done as described in section (2.4.3.1) and the results were arranged in a table.

2.5   THE KINETIC AND THERMODYNAMIC STUDIES OF LECTIN BINDING TO ERYTHROCYTE SURFACE GLYCOCONJUGATES USING MULTIPLE MYELOME SERUM SAMPLE

2.5.1 THE KINETIC STUDIES.

2.5.1.1 THE TIME-COURSE OF LECTIN BINDING TO ERYTHROCYTE SURFACE GLYCOCONJUGATES

Reagents

The standard erythrocyte suspension and the assay buffer (Tris-HCl pH8.5) containing 15 mM $CaCl_2$ and 0.2M of NaCl was prepared as described in section (2.4.1.1)

Procedure

1- At zero time, 0.5mL of erythrocyte suspension was added to 50µL of serum (3750µg protein), the final volume of the assay mixture was made up to 1mL with tris-HCl-buffer pH 8.5. The reaction mixture was incubated at 37°C for several time intervals (15, 30, 60, 90 and 120 minutes).

2- After each incubation time, the assay tubes were treated according to step 2 and 3 mentioned in section (2.4.1.1).

3- Parallel experiments were carried out according to section (2.4.1.2) to determine the amount of non-specific binding.

4- To determine the time-course of lectin binding to erythrocyte surface glycoconjugates at different temperatures, the above experiments were performed at (5, 15, 25 and 37°C).

Calculations

1- The concentration of lectin in milligrams per incubation medium (mg/mL) involved in total binding to erythrocyte surface glycoconjugates, was calculated according to the following formula:

$$\text{The concentration of lectin (mg/mL) involved in total binding} = \frac{A - A^*}{A} \times \text{The total concentration of lectin (mg/mL) used in the assay}$$

Where:

A: The absorbance of standard erythrocyte suspension at 620nm.

A^*: The absorbance of unbound (free) erythrocytes at 620nm.

2- The concentration of lectin in milligrams per incubation medium (mg/mL) involved in non-specific binding to erythroctye surface glycoconjugates, was calculated according to the following formula:

$$\text{The concentration of lectin (mg/ml) involved in non-specific binding} = \frac{A' - A^*}{A'} \times \text{total lectin concentration (mg/ml) used in the assay}$$

Where:

A': The absorbance of neuraninidase-treated erythrocytes at 620nm.

A^*: The absorbance of unbound (free) erythrocytes at 620nm.

3- The concentration of specifically bound lectin in milligrams per incubation medium (mg/mL) was calculated by subtracting the concentration of lectin involved in non-specific binding from the concentration of lectin involved in total binding:-

concentration of	concentration of		concentration of lectin
specifically bound =	lectin (mg/mL) .	-	(mg/mL) involved in
lectin (mg/mL)	involved in total binding		non-specific binding

The concentration of specifically bound lectin which represents the concentration of (lectin-glycocnjugate) complex, was expressed in micro-units (μU), since $1\mu U$ is the concentration of lectin (mg/mL) which gives 40% specific binding after incubation for 120 minutes at $37^{\circ}C$

4- The concentrations of specifically bound lectin (lectin-glycoconjugate) complex in micro-units were plotted against their corresponding incubation times.

2.5.1.2 DETERMINATION OF THE CONCENTRATION OF LECTIN BINDING SITES (Bmax) AND THE AFFINITY CONSTANT (Ka) OF LECTIN ASSOCIATION WITH ERYTHROCYTE SURFACE GLYCOCONJUGATES

Regents

The tris-HCl buffer pH 8.5 and the standard erythrocyte suspension were prepared as previously described in section (2.4.1.1).

Procedure

1- 0.5 mL of erythrocyte suspension was added to increasing amounts of lectin (0.75-3.75 mg/mL), the final volumes were made up 1mL with tris-HCl buffer (0.2 M, pH 8.5 contains 0.2M NaCl and 15 mM Ca Cl_2).

2- The assay tubes were incubated for 120 minutes at $37^{\circ}C$, then they were treated as mentioned in step 2 and 3 of section (2.4.1.1).

3- The previous steps were repeated at different temperatures: (5, 15, and $25^{\circ}C$).

Calculations

1- The concentration of specifically bound lectin (mg/mL) was calculated for each tube according to the calculations of section (2.5.1.1). The concentration was expressed in micro-units as described in section (2.5.1.1).

2- The concentration of free (unbound or unreacted) lectin was calculated by subtracting the concentration of lectin (mg/mL) involved in total binding from the total concentration of lectin (mg/mL) used in each experiments.

free lectin = Total amount of lectin – amount of lectin gives total binding.

The concentration of free lectin (mg/mL)	=	Total concentration of lectin (mg/mL)	-	The concentration of lectin (mg/mL) gives total binding

3- The concentration of lectin binding sites (B_{max}) and the affinity constant (Ka) were determined according to Scatchard equation[118].

$$\frac{B}{F} = \frac{1}{Kd} \times (B_{max} - B)$$

$$Ka = \frac{1}{Kd}$$

Where

B: The concentration of specifically bound lectin.

F: The concentration of free lectin.

Ka: The affinity constant.

Bmax: The maximal binding capacity.

Kd: The dissociation constant.

4- The plot of B/F values against the values of B, gives linear relationship. The total concentration of lectin binding sites (B_{max}) was calculated from the intercept on the x-axis, while the value of affinity constant was calculated from the slope of the straight line.

5- The Ka and (B_{max}) values were also determined from the Eadie-Hofstee plot of data getting from Scatchard plots, using the following equations.

$$B = -Kd \frac{B}{F} + B_{max}$$

The values of Ka and (B_{max}) were calculated from the slope of the straight line and the intercept on Y-axis respectively.

2..5.1.3 *DETERMINATION OF HILL-COEFFICIENT (n) OF LECTIN BINDING TO GLYCOCONJUGATES.*

Calculations

1- All data were obtained from the experiment mentioned in section (2.5.1.2).

2- The value of Hill coefficient (n) was calculated according to Hill equation.

$$\log(\frac{B}{B_{max} - B)} = n\log F - \log Kd$$

3- The values of log (B/(B_{max}-B)) was plotted against the values of logF, the Hill coefficient (n) was calculated from the slope of the straight line.

2.5.2 THE THERMODYNAMIC OF LECTIN BINDING TO ERYTHROCYTE SURFACE GLYCOCONJUGATES

Procedure

The same steps mentioned in section (2.5.1.1) and section (2.5.1.2) were performed

Calculations

1- The thermodynamic parameters of standard state were obtained from Van't Hoff plot, the values of the natural logarithm of affinity constant (Ka) obtained at different temperatures were plotted against the reciprocal values of the absolute temperature in Kelvin (1/T), according to the following equation:

$$\ln Ka = \frac{\Delta S^{\circ}}{R} - \frac{\Delta H^{\circ}}{RT}$$

Where:

ΔH°: The enthalpy change of the standard state.

ΔS°: The entropy change of the standard state.

R: The gas constant (8.31441 J.K^{-1} mol^{-1}).

ΔH° value was obtained from the slope of the linear relationship of the plot.

The change in Gibbs free energy of the standard state (ΔG°) was obtained from the following equation:

$$\Delta G^{\circ} = -RT \ln Ka$$

While the entropy change of the standard state ΔS° was obtained from:

$$\Delta S^{\circ} = \frac{\Delta H^{\circ} - \Delta G^{\circ}}{T}$$

2- The thermodynamic parameters of the transition state were obtained from Arrhenius plot of $\ln K_{+1}$ values against 1/T values, that gives a linear relationship according to the following equation:

$$\ln K_{+1} = \ln A - (\frac{Ea}{RT})$$

Where:

A: Arrhenius constant.

The value of apparent energy of activation (Ea) of the binding reaction can be determined from the slope of the straight line. The enthalpy of the transition state ΔH^* was obtained from:

$$\Delta H^* = Ea - RT$$

The free energy change of the transition state ΔG^* is calculated from the following equation:-

$$\Delta G^* = -RT \ln K_{=1} + RT \ln(\frac{KT}{h})$$

Where:

K: is Boltzmann constant (1.38×10^{-23} Jdeg^{-1})

h: is Plank constant (0.662×10^{-33} JS^{-1})

The change in entropy of the transition state ΔS^* is calculated from the following formula:

$$\Delta S^* = \frac{\Delta H^* - \Delta G^*}{T}$$

CHAPTER THREE

Results

&

Discussion

3.1 DETERMINATION OF DIFFERENT BIOCHEMICAL CONSTITUENTS IN SERA OF PATIENTS WITH MULTIPLE MYELOMA, LEUKEMIA AND BREAST CANCER.

3.1.1 DETERMINATION OF TOTAL PROTEIN(TP) AND TOTAL SIALIC ACID (TSA) CONCENTRATIONS IN SERA OF PATIENTS WITH MULTIPLE MYELOMA, LEUKEMIA, AND BREAST CANCER.

Serum samples from seventy two persons were tested for total protein (TP) and total sialic acid (TSA). These included thirty samples from normal individuals, twenty two samples from patients with multiple myeloma, twenty five samples from patients with leukemia, and twenty five samples from patients with breast cancer.

The mean concentrations ± SD of TP and TSA in sera of normal controls and cancer patients are summarized in table (3-1).

The serum levels of total protein were measured using the method of Lowry et. al.[111]

The results in table (3-1) reveal that the means of total protien concentrations are slightly changed from one type of cancer to another, no significant differences have been observed between the three types of cancer diseases, but in general increased TP levels were reported in cancer patients in comparison with healthy group (P<0.01).

The increased total protein levels in cancer patients may be an evidence for the uncontrolled behavior of the cancerous cells. for instance, the elevation of total protein levels in multiple myeloma reflects the excessive production of paraproteins by cancerous plasma cells[101].

The results presented in table (3-1) show also an increased total sialic acid (TSA) levels in all individuals with cancer diseases when compared to normal individuals.

There was a highly significant difference between the serum total sialic acid in multiple myeloma patients compared to the control group; 105.63±25.23 mg/dL *versus* 63.8±9.3 mg/dL respectively (P<0.001). Also the total sialic acid (TSA) levels were found to be significantly elevated in sera of leukemia and breast cancer patients, 100.42±23.8 mg/dL, 100.18±16.42 mg/dL respectively (P<0.001), when compared to control group. No significant differences of serum TSA levels in the three groups of cancer patients were observed.

Table (3-2) shows the specificity and sensitivity of TSA test. The specificity values were calculated as the number of cases having TSA levels less or equal the upper limit of normal (73 mg/dL) divided by the total number of cases. The sensitivity of TSA test was determined by taking the number of cases which have TSA levels more than the upper limit (73 mg/dL) divided by the total number of cases.

In general, the sensitivity of TSA assay varies from 20% to 92%, two out of twenty two patients with multiple myeloma had TSA values within the normal range exhibiting high sensitivity (90.9 %) of the assay, the sensitivities for TSA measurements in leukemia and breast cancer patients were 84% and 92% respectively. TSA specificity in all groups involved in this study were lower than the corresponding sensitivities, they were 9%, 16%, and 8% in myeloma, leukemia and breast cancer respectively.

Many glycoproteins and glycolipids from neoplastic cells differ in carbohydrate composition from those found in normal cells, since many of these glycoconjugates contain terminal sialic acid which can be shed into the circulation[7].

The potential role of sialic acid in the mechanism of tumor formation is indicated by the finding that sialic acid mask the surface of certain tumor cells by interfering with the immune response of the host [119], and that sialic acid content appears to be correlated with metastatic ability in a variety of tumor cells [53].

Total sialic acid (TSA) is of great interest as a marker of malignancy although it has not been demonstrated to be specific for any type of cancer. Serum TSA levels have been found to be elevated in a number of different cancers [38,39,41,44].

In addition, serum total sialic acid has been reported to be increased in chronic liver diseases, pneumonia, and rheumatoid arthritis. Further, patients with chronic glomerulonephritis also have elevated serum TSA concentrations [110]. It can only be said that increase in serum TSA is associated with certain diseases, including cancer, and is roughly related to tumor size.[7].

Gail et. al [120] have reported on the utility of TSA levels for the diagnosis of lung cancer as well as for discriminating advanced lung cancer. TSA level were also correlated with different histologic types in cancer patients [47, 121].

Erbil et al. [122] reported on increased levels of TSA in genitourinary tumors, concluding that serum TSA levels were highly correlated with the stage and grade in patients with advanced urological cancer.

Our results show that patients with three types of cancers had high serum TSA levels and there is a significant elevation in TSA ($p<0.001$) when compared to normal group. Figure (3-1) shows the distribution of the individual values of TSA in sera of cancer patients and normal persons. Multiple myeloma patients show more elevated values of TSA compared with normal subjects.

Also the mean TSA concentrations in multiple myeloma is higher when compared with the values obtained in leukemia and breast cancer patients, the differences were not significant. The sensitivity values indicate that serum total sialic acid test is sensitive to cancers but less specific.

The elevation of TSA concentrations in sera of cancer patients can be attributed to the increased turnover of the highly sialylated cancerous cells and also to the shedding of sialic acid residues at the non-reducing end of glycoconjugates on the cell surfaces. In addition, the excessive production of immunoglobulin-derived proteins in multiple myeloma patients may results in elevated total sialic acid levels as the immunoglobulins are richly sialylated [110].

The individual values of TSA levels in all groups were normalized to the corresponding TP levels. The index (TSA/TP) showed significant differences (P<0.01) between each group of cancer patients and normal subjects, the results are shown in table (3-1).

It has been well established in this investigation that serum total sialic acid concentrations were considerably higher in cancer patients than in control subjects. This observation is in keeping with those of other studies which have reported an elevation of serum total sialic acid in other malignant conditions. Total sialic acid measurements appeared to have low specificity but high sensitivity for cancer diseases. Serum TSA may be of value as a tumor marker but prospective studies are required.

Table (3-1) : Serum total sialic acid and total protein levels in patients with multiple myeloma, leukemia and breast cancer. Details are described in section (2.2.1) and (2.2.2)

Group	No.	TSA mg/dL ±SD	TP g/dL ± SD	TSA/ TP mg/g ±SD
Multiple myeloma	22	105.63 ± 25.23	7.8 ±1.03	14.11± 1.66
Leukemia	25	100.4 ±23.8	7.6 ±1.06	15.32 ±2.2
Breast cancer	25	100.18±16.42	7.1 ±1.02	14.23 ±1.45
Control	30	63.8±9.3	7.03 ±0.83	9.04 ±2.02

Table (3-2) : Specificity and Sensitivity of TSA measurement

Group	No.	Specificity* %	Sensitivity** %
Multiple myeloma	22	9	90.9
Leukemia	25	16	84
Breast cancer	25	8	92
Control	30	86	13

* The number of cases have TSA values ≤ 73 mg/dL divided by the total number of cases by 100.

** The number of cases have TSA values > 73 mg/dL divided by the total number of cases by 100.

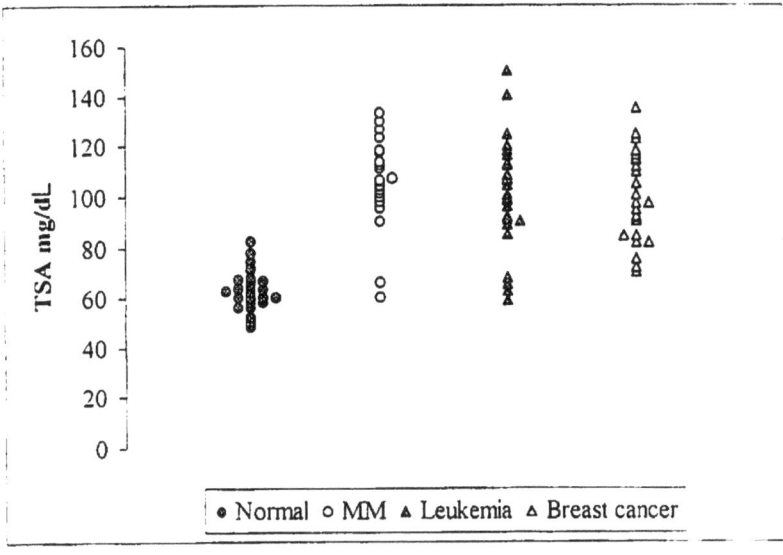

Figure (3-1): The distribution of the individual values of TSA in multiple myeloma, leukemia, and breast cancer. Details are described in section (2.2.2).

3.1.2 DETERMINATION OF LIPID-BOUND SIALIC ACID (LBSA) LEVELS IN SERA OF PATIENTS WITH MULTIPLE MYELOMA, LEUKEMIA AND BREAST CANCER.

Serum lipid–bound sialic acid (LBSA) levels were determined in healthy persons and in patients with multiple myeloma, leukemia, and breast cancer using the method of Katopodis et.al[112]. This method is widely used for the measurement of LBSA because of its rapidity, good precision, and low cost.

Table (3-3) shows the mean values of LBSA concentrations in three groups of cancer patients and in normal individuals. The results presented in this table reveal an overall elevation in LBSA levels for each group of patients with cancer when compared to normal healthy individuals. The increased LBSA levels in sera of patients with multiple myeloma showed statistically significant differences when compared to those values obtained from sera of normal individuals, the mean concentrations of LBSA in multiple myeloma and in normals were 34.06 ± 11.64 mg/dL *versus* 16.64 ± 3.2 mg/dL ($P<0.001$). Also high LBSA concentrations were observed in leukemia and breast cancer patients. The mean values of LBSA in leukemia and breast cancer were 31.6 ± 10.13 mg/dL and 30.04 ± 12.23 mg/dL respectively, and were found to be statistically significant when compared to the control group ($P<0.001$). Only insignificant differences were observed to be present in three groups of cancer diseases.

The percentages of LBSA test specificity and sensitivity were calculated using the value 19.84 mg/dL as the upper limit of normal.

In general, high sensitivity was observed, since the sensitivity percents were 95%, 100%, and 88% in multiple myeloma, leukemia, and breast cancer respectively. Test specificity was lower, since it ranged between zero and 12%. Among the substances that are shed from growing malignant cell gangliosides, in particular lipid–bound sialic acid (LBSA) showed significant elevations in sera from patients bearing cancer. Thus, serum LBSA has been proposed as a useful tumor marker, more satisfactory than total sialic acid[123, 124].

Lipid- bound sialic acid is expected to behave as a non-specific tumor marker after shedding, and that is an important aspect of the turnover of normal cell surface constituents which occurs principally in both normal and malignant growing cells [125].

The present investigation showed significantly elevated levels of LBSA in cancer patients when compared with controls ($P<0.001$), this finding is in agreement with other studies which have reported an elevation of serum LBSA in other cancer diseases [125, 126].

Dnistrian M. and Schwartz K. [127], measured the LBSA levels in seven different cancer cases, the results were compared with carcinoembryonic antigen (CEA) levels; it has been shown that LBSA is a useful marker in cancer, particularly in patients with leukemias, lymphomas, Hodgkin's disease, and melanomas. It has also shown that LBSA was a valuable marker in lung cancer especially when combined with CEA, and it may be a suitable marker in colon cancer, LBSA was found to be less useful in breast cancer [127].

In a recent study [59], it has been reported that there is a variation in LBSA serum levels related to disease extent, i.e., the more advanced the disease, the higher is the LBSA serum value, in the following order: local < locoregional < metastatic disease. This finding can be taken as a proof that LBSA can be used as a marker for the assessment of disease extent.

Vivas et.al [128] conclude that the measurement of LBSA appears to be of no value either for the early detection of cervix cancer, or as a complement in the clinical staging of this tumor.

Erbil et. al. [124] investigated LBSA as a marker of colorectal cancer and concluded that it was useful for differential diagnosis and disease monitoring, but not for the early diagnosis of these tumors.

Polivkova et. al. [48] believe that the determination of LBSA levels could be useful not only for cancer diagnosis but also prognosis.

Toumbis et. al. [129] did not find any significant differences in LBSA levels between benign and malignant pleural effusions.

Some authors [123.130] have based their work upon the fact that alterations in glycolipid metabolism are well documented in many tumors, including human cancers. These authors have reported raised levels of LBSA in sera of patients with various neoplasms, but others have found raised LBSA values in sera of patients with acute inflammatory diseases. This has led to the conclusion that the high LBSA levels in sera of patients with cancer probably emerge from associated inflammation and, hence, the acute phase reactant glycoproteins mainly build up the LBSA fraction, and this may reflect the low specificity of the marker.

From our results, LBSA appears to be useful marker in distinguishing between healthy individuals and cancer patients, since LBSA exhibits a high sensitivity. The combined use of LBSA with other markers may provide high degree of marker positivity.

Table (3-3): Serum Lipid-bound sialic acid (LBSA) levels in sera of patients with multiple myeloma, leukemia and breast cancer. Details are described in section (2.2.3).

Group	No.	LBSA mg/dL ± SD	Specificity% *	Sensitivity% **
Multiple myeloma	22	34.06±11.64	4.5	95
Leukemia	25	31.6±10.13	-	100
Breast cancer	25	30.04±12.23	12	88
Control	30	16.64±3.2	86	13

* Specificity was calculated by division the number of cases having LBSA values ≤ 19.84 mg/dL by the total number of cases by 100.

** Sensitivity was calculated by division of the number of cases having LBSA>19.84 mg/dL by the total number of cases by 100.

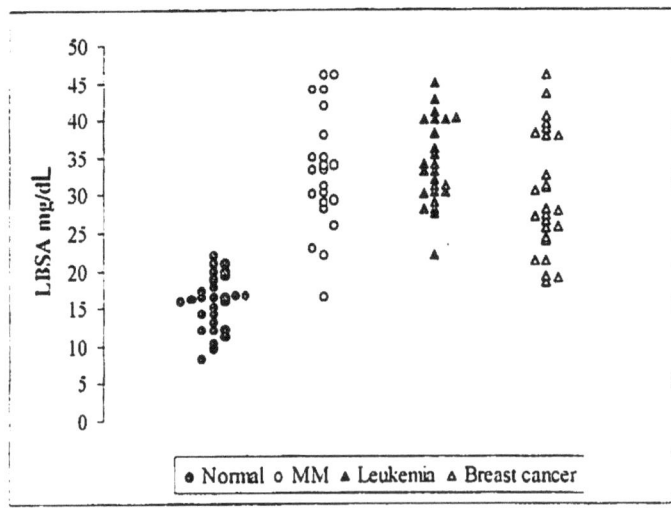

Figure (3-2): Distribution of the individual values of LBSA in multiple myeloma, leukemia and breast Cancer.

3.1.3 DETERMINATION OF MUCOID PROTEINS (MP) AND PROTEIN-BOUND HEXOSES (MANNOSE+GALACTOSE) IN SERA OF PATIENTS WITH MULTIPLE MYELOMA, LEUKEMIA AND BREAST CANCER.

The levels of mucoid proteins (MP) and protein-bound hexoses were determined in sera of multiple myeloma, leukemia and breast cancer patients using the method of Weimer and Mashin [113].

The mean concentrations of mucoid proteins and protein-bound hexoses in sera of all patients and normal healthy controls are summarized in table (3-4) and table (3-5) respectively. Figure (3-3) and figure (3-4) shows the distribution of the individual values of mucoid proteins and protein-bound hexoses respectively.

From our results, statistically significant elevations were observed in the serum levels of mucoid proteins in cancer patients when compared to normal persons (P<0.001). The mean concentrations of muciod protiens (MP) reached to 17.95±1.12 mg/dL in multiple myeloma patients, whereas in leukemia patients it was 17.32±1.07 mg/dL and 17.56±1.18 mg/dL in breast cancer patients. Only minor differences in MP concentrations were observed when the three groups of cancer were compared to each other.

The values of specificity (true negative) and sensitivity (true positive) for the MP test are shown in table (3-4). Test specificity and sensitivity were calculated considering the value 11.46 mg/dL as the upper limit of normal. Test specificity ranged between 12% and 16%. Only three from twenty five patients with breast cancer had MP levels within the normal range giving 88% sensitivity, test sensitivity in multiple myeloma and leukemia were 86% and 84% respectively.

As can be seen from table (3-5), all patients exhibit higher levels of serum protein-bound hexoses. By employing Student's t-test, statistically significant differences (P<0.01) were found to be present between each group of cancer patients and normal group, but only insignificant changes were found between cancer diseases.

The upper limit for protein- bound hexoses of normal was 118.01 mg/dL and was used for comparing the other values of protein-bound hexoses in patients to calculated the test specificity and sensitivity. The results obtained revealed low specificity since it ranged between 9% and 12%, on the contrary, test sensitivity showed high percentages, 90%, 88% and 92% in multiple myeloma, leukemia and breast cancer respectively.

The present investigation was carried out to clarify the possible usefulness of serum mucoid proteins (MP)and protein-bound hexoses as biomarkers for identifying cancer diseases. The mucoid proteins and carbohydrate – rich fraction of proteins has been studied as an indicator of tumor presence and the changes in glycoproteins could be determined by their carbohydrate moities[56]. Many investigators have demonstrated elevated levels of circulating glycoproteins in cancer patients. Previous reports have also stated that the increased levels of mucoid proteins (MP) and serum hexoses (galactose + mannose) bear a positive correlation in malignancies with stage of diseases[131].

From our results, it is apparent that the mean levels of both mucoid proteins and protein-bound hexoses are significantly higher in cancer patients than in normal healthy controls (P<0.001) (P<0.01) respectively, table (3-4) and table (3-5). These findings is in accordance with those of other investigators[131].

Patel, P.S. et al[132] has observed significant increase in the levels of the two biomarkers in breast carcinoma patients compared with the normal controls, also the differences were significant when compared to the patients with benign breast diseases, it has also been suggested that the measurement of the two markers be helpful in the diagnosis of breast carcinoma as well as in differentiating between lobular carcinoma and infilterating duct carcinoma patients.

Also the levels of mucoid proteins and protein-bound hexsoses were evaluated in lung cancer patients and patients with benign lung diseases, it was revealed that the levels of the two markers were significantly elevated in lung cancer compared to benign lung diseases and normal individuals. In addition, comparison between patients with benign lung diseases and normal revealed a significant differences in the levels of protein-bound hexoses, whereas no significant differences in mucoid protein levels, and it was suggested that the measurement of the markers may be useful in the diagnosis and histological typing of lung cancer[56].

Breadly et. al. [133] have examined the possibilities of the source of increased protein-bound carbohydrates occurring in cancer and reported that there may be glycoprotein synthesis by tumor itself.

In a recent work, Bhuvarahamurthy, V. et al.[61] have shown that serum glycoproteins and glycosidases were elevated in patients with different stages of cervical cancer, and they have suggested that the increase in circulating glycoproteins may be due to increased activities of glycosidases and, hence, increased rates of membrane glycoprotein degradation and shedding of these excessive glycoproteins into the sera.

The quest is ongoing for more reliable serum markers for detecting malignant diseases. Because of the crucial role of cell surface and membrane constituents in neoplastic behaviour, changes in normal glycoconjugates have long been associated with malignancies[132].

In agreement with other observations, our investigation indicates that the levels of glycoproteins were elevated in cancer patients and that the mucoid proteins test and protein-bound hexoses test may be of clinical importance.

These observations can be adopted as a useful parameters for diagnosis and follow up investigations, but further investigations on a large scale are required to establish the value of these glycoproteins to be used as a malignancy markers. Mucoid proteins test exhibits more specificity when compared to the low specific protein-bound hexoses test, from this it seems rationally that MP test is more useful in discrimination between cancer patients and normal individuals.

Table(3-4): Serum levels of mucoid proteins (MP)in patients with multiple myeloma, leukemia and breast cancer, and the test specificity and sensitivity. Details are described in section (2.2.4.1).

Group	No.	MP mg/dL ± S.E.	Specificity% *	Sensitivity% **
Multiple myeloma	22	17.95±1.12	13	86
Leukemia	25	17.32 ± 1.07	16	84
Breast cancer	25	17.56±1.18	12	88
Control	30	10.26±1.2	90	10

* Calculated as the number of cases having MP≤11.46 mg/dL divided by the total number of cases by 100.

** Calculated as the number cases having MP> 11.46 mg/dL divided by the total number of cases by 100.

Table (3-5): Serum levels of protein-bound hexoses in multiple myeloma, leukemia, and breast cancer and the test specificity and sensitivity. Details are described in section (2.2.4.2).

Group	No.	Protein-bound hexoses mg/dL ± S.E.	Specificity% *	Sensitivity% **
Multiple myeloma	22	136.53±3.71	9	90
Leukemia	25	137.82±3.09	12	88
Breast cancer	25	134.62±4.11	8	92
Control	30	112.2±5.21	83	16

* Calculated as the number of case have protein-bond hexoses ≤118.01 mg/dL divided by the total number of cases by 100.

** Calculated as the number of case have protein-bond hexoses >118.01 mg/dL divided by the total number of cases by 100.

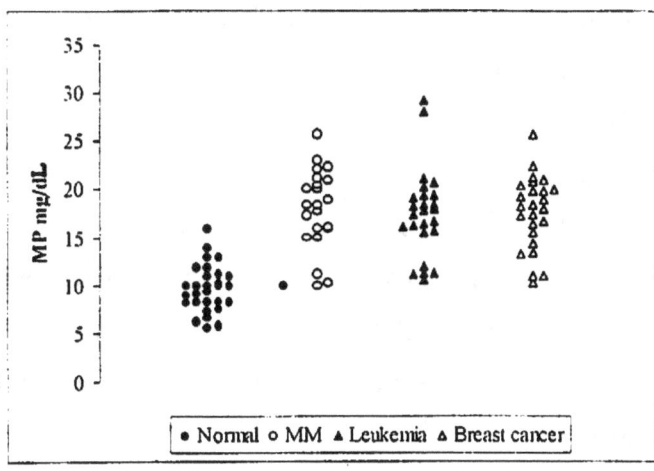

Figure(3-3): Distribution of the individual values of MP in sera of patients with multiple myeloma, leukemia and breast cancer. Details are described in section (2.2.4.1).

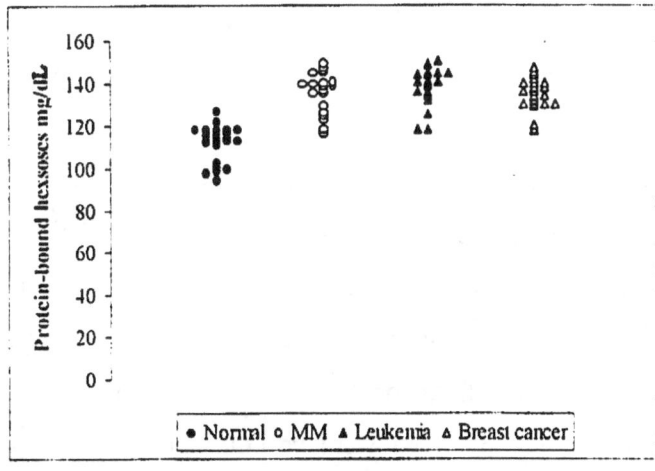

Figure(3-4): Distribution of the individual values of protein-bound hexoses in sera of cancer patients. Details are described in section (2.2.4.2).

3.2 DETERMINATION OF SUPEROXIDE DISMUTASE ACTIVITY IN SERA OF PATIENTS WITH MULTIPLE MYELOMA, LEUKEMIA AND BREAST CANCER.

Superoxide dismutase activity was determined in sera of 14 multiple myeloma patients, 12 leukemia patients,15 breast cancer patients and 10 normal controls. SOD activity was measured by the method of Winterbourn et al [114], the method is based on the ability of the enzyme to inhibit the reduction of nitroblue tetrazolium (NBT) by superoxide generated during the reaction of photoreduced riboflavin and oxygen [66].

Table (3-6) shows the mean of SOD activity in normal donors and the three groups of cancer patients. The results revealed that cancer patients had low values of SOD activity compared with normal individuals (p<0.001). The mean values of SOD in sera of multiple myeloma, leukemia, and breast cancer patients were 1.32 ± 0.13, 1.36 ± 0.18 and 1.42 ± 0.14 respectively. No reliable differences were found in the SOD activity between the different cancer diseases. Figure (3-5) shows the distribution of the individual values of SOD activity in cancer patients.

It has been reported that there is a relationship between SOD activity and cancer, since decreased amounts of manganese- containing SOD (Mn SOD) have been found in spontaneous, transplanted, virally induced, in vitro and in vivo tumor cells, also lowered amounts of copper – zinc – containing SOD have been found in many, but not all, tumors [62]. With comparison to normal cells, malignant cells were found to have lower levels of SOD activity, in addition. Fernandez – pol et al [134], has found that high metastatic cell lines contain less SOD than low metastatic cell ones.

Our results show lower SOD activity in sera of all patients with cancer diseases when compared to normal persons, these results are in agreement with the reports of other investigators.

Tumor cells have been shown to produce superoxide free radicals, if the rate of production of superoxide ion in tumor mitochondria is comparable to that found in the mitochondria from normal tissue, then the loss of SOD would results in a net increase in the level of superoxide ion in the tumor cell [62]. Differences in the serum activity of SOD in patients with cancer diseases are probably due to the decreased enzyme level in the tumor cell. Another possibility is a decrease in the synthesis and release of SOD from the blood cells because of the malignant immune deficiency [66]. Such deficiency appears, for example,when enhanced serum levels of gangliosides (e.g. LBSA) are released by the tumor cells. LBSA binds to the plasma membranes of the mononuclear cells and inhibits their functions, which may be an important mechanism for immunosuppression in malignant diseases [66].

The values of serum SOD activity obtained from this assay were correlated with LBSA concentrations in sera of multiple myeloma, leukemia and breast cancer patients. The results disclosed that there is a significant negative correlation between the two parameters. Consequently, the ratio of LBSA/SOD was calculated for each patient. Table (3-6) shows the values of LBSA/SOD ratio for the three groups of cancer patients and the healthy individuals. The LBSA/SOD ratio was about threefold higher in multiple myeloma patients compared with normal healthy individuals 27.54±9.32 *versus* 9.89±7.17 (P<0.001). The ratio of LBSA/SOD was 23.5±6.78 in leukemia patients and was significantly higher than in normal group (P<0.001), also the ratio was higher (22.62±7.42) in breast cancer patients when compared with normal group. No significant differences were found in the three groups of cancer diseases.

Serum levels of SOD and LBSA reflect the changes in the content of membrane glycolipids and cellular activity of SOD. Some authors admit that these changes in the tumor cell are in closed connection; the membrane of the neoplastic cell has an altered lipid content and structure organization, which leads to decreased antioxidant protection. The loss of cell differentiation leads to an increase of the cell glycolipids and, on the other hand, to a decrease in the intracellular SOD [135]. Our results show that in the serum from patients with cancer diseases such dependence exists, since patients with elevated LBSA levels have low SOD activities in the serum figure (3-6). It has been reported that such negative correlation does not appear in children with non-cancer diseases. This is in accordance with the observations that changes in membrane glycolipids and cellular antioxidants occur in malignant, but not benign, tumors[66]. These findings suggest that SOD and LBSA are good tumor markers.

Table (3-6): Serum SOD activity, LBSA and LBSA/SOD ratios in patients with multiple myeloma, leukemia and breast cancer. Details are described in section (2-3)

Group	No.	LBSA mg/dL (mean±SD)	SOD activity (mean±SD)	LBSA/SOD (mean±SD)
Multiple myeloma	14	33.85±12.31	1.34±0.13	27.54±9.32
Leukemia	12	31.01±11.43	1.36±0.18	23.5±6.78
Breast cancer	15	30.33±10.01	1.42±0.14	22.62±7.42
Control	10	16.64±3.2	1.85±0.17	9.89±7.17

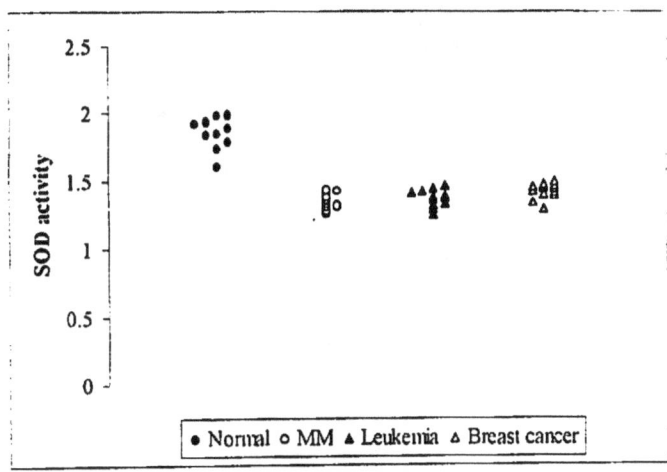

Figure(3-5): Distribution of the individual values of SOD activity in sera of patients with multiple myeloma, leukemia and breast cancer. Details are described in section (2.3)

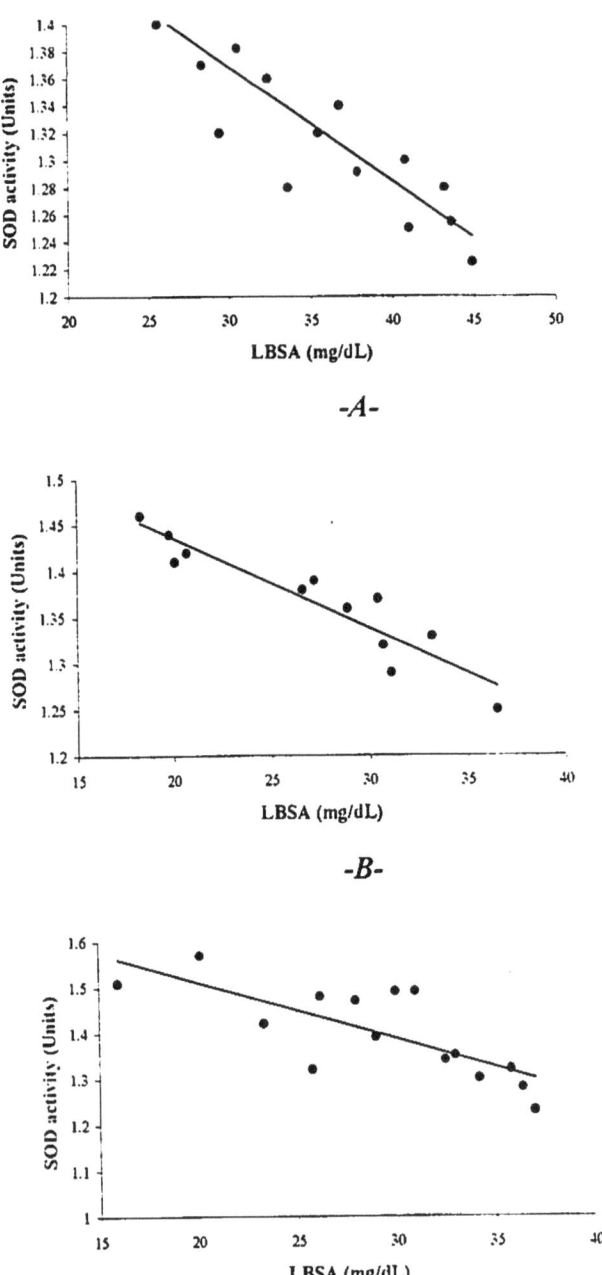

Figure(3-6): Correlation between SOD activity and the levels of LBSA in
sera of patients with cancers.
-A- Multiple myeloma.
-B- Leukemia
-C-Breast cancer

3.3 BINDING STUDIES OF LECTIN TO ERYTHROCYTE SURFACE GLYCOCONGUGATES IN MULTIPLE MYELOMA AND LEUKEMIA SERUM SAMPLES

3.3.1 PRELIMINARY TEST OF LECTIN BINDING TO ERYTHROCYTE SURFACE GLYCOCONJUGATES.

The ability of lectin to bind the carbohydrate residues of erythrocyte surface glycoconjugates was tested using two different serum samples as a source of cancerous lectin, one of the samples belongs to multiple myeloma patient and the other to leukemia patient. Human blood type (A) was used in all experiments of binding studies as a source for glycoconjugates to which cancerous lectin will bind.

The total binding of lectin to glycoconjugates was estimated according to the hemagglutination assay [116-117]. Hemagglutination assay is a semiquantitative procedure and has been widely used as a laboratory test because of its ease and versatility. It depends on aggregating and sedimentation of the erythrocytes after reaction with the bivalent or multivalent lectin [136]. The hemagglutination unit (H.U.) is the amount of lectin which reduce the absorbance of standard erythrocyte suspension (2 at 620 nm) to 50%.

Non-specific binding was tested using neuraminidase to be incubated with the erythrocyte suspension before the assay, this enzyme is responsible for the release of terminal sialic acid residues from the erythrocyte surface glycoconjugates, and hence, the penultimate N-acetylgalactosamine will be exposed for the lectin binding.

Many factors may influence the hemagglutination assay these include lectin concentration, particle charge, pH, ionic strength, temperature and the time of incubation [136].

3.3.2   FACTORS   EFFECTING   LECTIN   BINDING   TO
GLYCOCONJUGATES   ON   HUMAN   ERYTHROCYTE
SURFACE.

3.3.2.1 EFFECT OF LECTIN AMOUNT ON ITS BINDING TO
ERYTHROCYTE SURFACE GLYCOCONJUGATES.

To estimate the suitable amount of lectin, 500 μL of standard erythrocyte suspension were incubated with increasing amounts of serum (cancerous lectin) for 30 minutes at 25°C. The results revealed that the binding of lectin to glycoconjugates increases with increasing amount of lectin added. Figure (3-7) represents the binding of lectin to erythrocyte surface glycoconjugates. The figure shows that the using of excessive amounts of lectin more than 3750 μg has no effect on the percent of specific binding, this indicates that the reaction of binding reaches the equilibrium state at this amount of lectin. Accordingly, 50 μL of serum (3750 μg protein) was used in all the subsequent experiments, since it gives maximum binding.

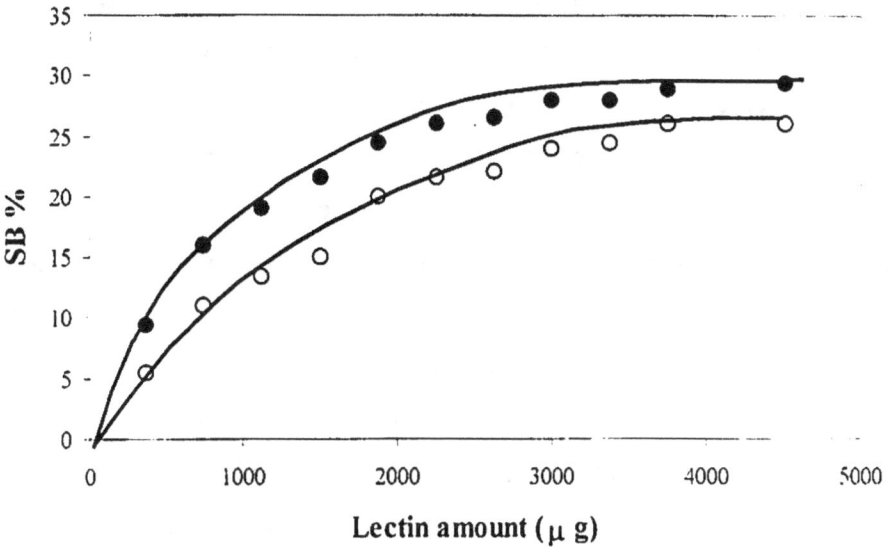

Figure (3-7): Effect of the amount of cancerous lectin on its binding to erythrocyte surface glycoconjugates.

• Multiple myeloma serum sample.

o Leukemia serum sample.

Details are described in section (2.4.2.1)

3.3.2.2 EFFECT OF pH ON LECTIN BINDING TO ERYTHROCYTE SURFACE GLYCOCONJUGATES

The effect of pH of the reaction medium on the binding of lectin to erythrocyte surface glycoconjugates was investigated using tris-HCl buffer solutions with pH values ranged from 7 to 9. Figure (3-8) shows the effect of pH on lectin binding, from this figure it is obvious that the optimum binding was obtained by incubating the erythrocyte suspension and cancerous lectin at pH=8.5. Therefore, the incubation in all subsequent experiments was carried out in the presence of tris-HCl buffer of pH 8.5. The results obtained from this assay indicate that the lectin binding is a pH-dependent process.

The observations that the percent of specific lectin binding decreases with the change of pH towards acidity or high basicity suggests that the abundance of H^+ ions in the acidic medium may inhibit the binding sites on both glycoconjugate and lectin molecules, OH^- ions in more basic medium may influence in the same manner, another possibility is that the changes in pH of incubation medium may results in denaturation of protein molecules involved in the binding and/or degradation of lectin-glycoconjugate complex. The decreasing in specific binding with increasing acidity also may be due to alterations in the chemical structure of lectin molecules.

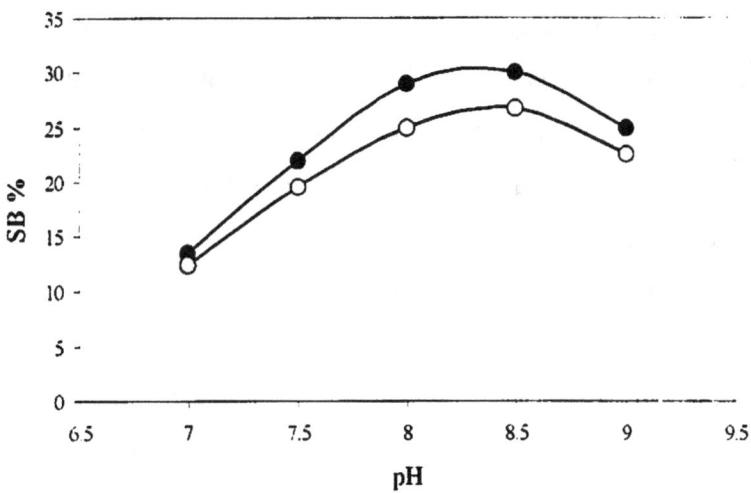

Figure (3-8): Effect of pH on lectin binding to glycoconjugates of erythrocyte surface using:

• Multiple myeloma serum sample
o Leukemia serum sample.

Details are described in section (2.4.2.2)

3.3.2.3 EFFECT OF TEMPERATURE ON THE BINDING OF LECTIN TO ERYTHROCYTE SURFACE GLYCOCONJUGATES.

The effect of different temperatures on the binding of cancerous lectin to erythrocyte surface glycoconjugates was investigated by incubating the erythrocyte suspension and cancerous lectin at different temperatures ranged between 5°C to 37°C. The temperature dependency of the binding is shown in figure (3-9) which revealed that there is an increase in the lectin binding with increasing temperature. The highest binding was achieved by incubating the erythrocyte suspension and the lectin at 37°C. Accordingly, in all subsequent experiments the temperature of 37°C was used as the incubation temperature. The results obtained from this assay show   that the lectin binding is a temperature – dependent process. The test was not achieved at any other temperature above 37°C, presumably because any other temperature may results in denaturation of the glycoconjugates or the lectin, since the lectin is a protein it is sensitive for the increased temperature, consequently, the binding will be less than that obtained at 37°C.

3.3.2.4 EFFECT OF INCUBATION TIME ON THE BINDING   OF LECTIN TO ERYTHROCYTE SURFACE GLYCOCONJUGATES.

To estimate the optimum incubation time, the erythrocyte suspension and cancerous lectin were incubated for different times ranged between 15 minutes and 150 minutes. Figure (3-10) demonstrates the effect of incubation time on lectin binding. The results obtained in this assay shows that the incubation for 120 minutes gives the highest binding. Consequently, in all subsequent experiments the incubation time was 120 minutes.

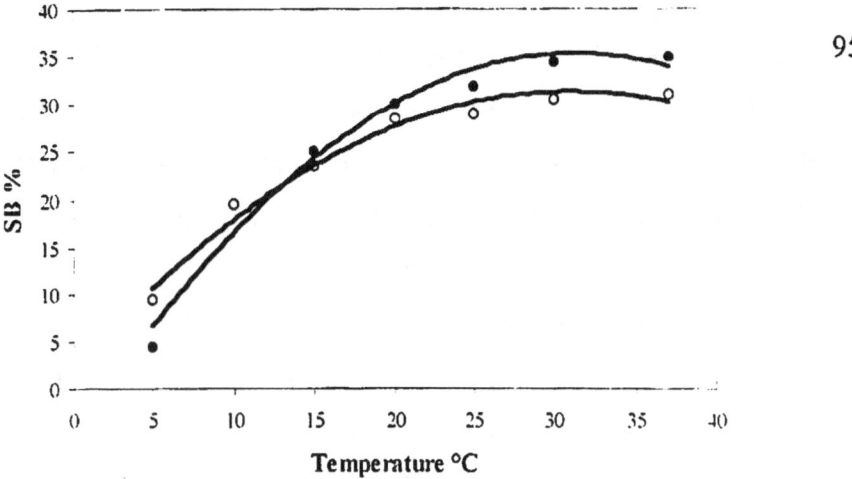

Figure (3-9): Effect of temperature on the binding of lectin to erythrocyte surface glycoconjugates using:

  • Multiple myeloma serum sample.

  o Leukemia serum sample.

Details are described in section (2.4.2.3)

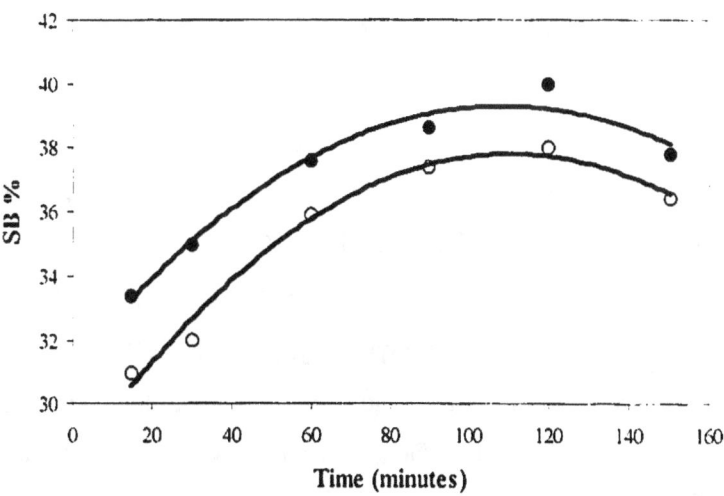

Figure (3-10): Effect of incubation time on lectin binding to erythrocyte surface glycoconjuates. using:

  •Multiple myeloma serum sample

  oLeukemia serum sample.

Details are described in section (2.4.2.4)

3.3.2.5 EFFECT OF Ca^{+2} IONS CONCENTRATION ON THE BINDING OF LECTIN TO ERYTHROCYTE SURFACE GLYCOCONJUGATES.

The effect of Ca^{+2} ions concentration on the binding of lectin to erythrocyte surface glycoconjugates was investigated using buffer solutions containing different concentrations of Ca^{+2} ions. Figure (3-11) shows the effect of Ca^{+2} concentration on lectin binding. It seemed that the binding of lectin to glycoconjugates was affected by Ca^{+2} ions, and the results obtained shows that the highest binding can be achieved using 15mM of Ca^{+2}, also the same concentration was used in all subsequent experiments. The results established that the lectin involved in this assay is a Ca^{+2} – dependent, and it can be said that no notable binding can be seen in the absence of Ca^{+2} ions. The possible role for Ca^{-2} ions in the binding of lectin to glycoconjugates is that it can stabilize the lectin –glycoconjugate complex since the binding of Ca^{+2} ions may induce conformational changes in the structure of the protein. Different Ca^{-2} – dependent lectins have been purified from various sources and most of these possess multimeric sturctures and are capable of forming cross – linked complexes[67].

3.3.2.6 EFFECT OF IONIC STRENGTH ON THE BINDING OF LECTIN TO ERYTHROCYTE SURFACE GLYCOCONJUGATES.

In general, the ionic strength of the incubation medium is thought to have a marked influence on lectin binding to glycoconjugates. The effect of NaCl and $MgCl_2$ were investigated. The results showed that the highest binding was obtained in the presence of 200mM NaCl. Consequently this concentration was used in all subsequent experiments. The results showed that different $MgCl_2$ concentrations have low effect on lectin binding. Figures (3-12A) and (3-12B) show the effect of NaCl and $MgCl_2$ respectively on binding.

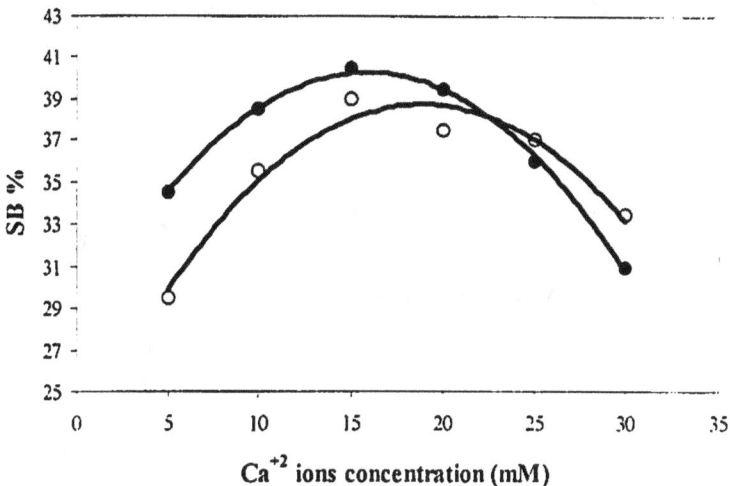

Figure (3-11): Effect of Ca^{+2} ion concentration on the binding of lectin to erythrocyte surface glycoconjugates using:

•**Multiple myeloma serum sample**

o **Leukemia serum sample.**

Details are described in section (2.4.2.5)

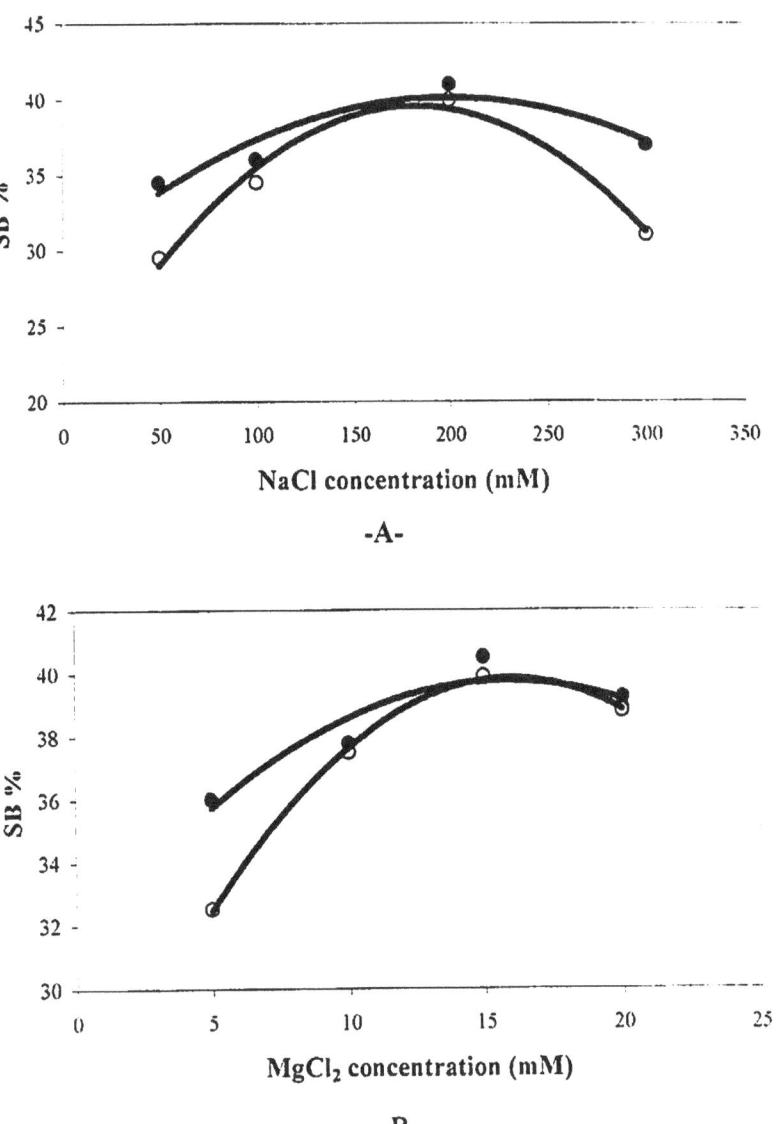

-A-

-B-

Figure (3-12): The effect of ionic strenght on lectin binding to erythrocyte surface glyconconjugates.

•Multiple myeloma serum sample

o Leukemia serum sample.

A: Effect of NaCl

B: Effect of $MgCl_2$

Details are described in section (2.4.2.6)

3.3.2.7 EFFECT OF DIFFERENT DENATURATING AGENTS ON THE BINDING OF LECTIN TO ERYTHROCYTE SURFACE GLYCOCONJUGATES.

Different denaturating agents were used to investigate their effect on lectin binding to glycoconjugates. Table (3-7) demonstrates the effect of 0.15 M of NaOH and HCl on the binding. The results revealed that NaOH and HCl considerably reduce the percent of specific binding, their denaturating effect is due to great changes in pH of incubation medium.

The effect of urea on lectin binding was investigated using different concentrations ranged between 3M and 6M. Figure (3-13 A) shows that the percent of specific binding decreased with increasing urea concentrations, this effect can be attributed to the effect of urea on the hydrophobic forces between protein molecules.

Figure (3-13B) shows the effect of different concentrations of PEG on the binding. Increasing concentrations of PEG may results in precipitation of protein molecules which leads to decrease the interaction between lectin and glycoconjugates, and hence decrease the percent of specific binding.

Table (3-7): Effect of 0.15M NaOH and HCl on lectin binding to erythrocyte surface glycoconjugates. Details are described in section (2.4.2.7)

Denaturating agent	SB% (MM)	SB% (leukemia)
NaOH	3	4
HCl	2	2

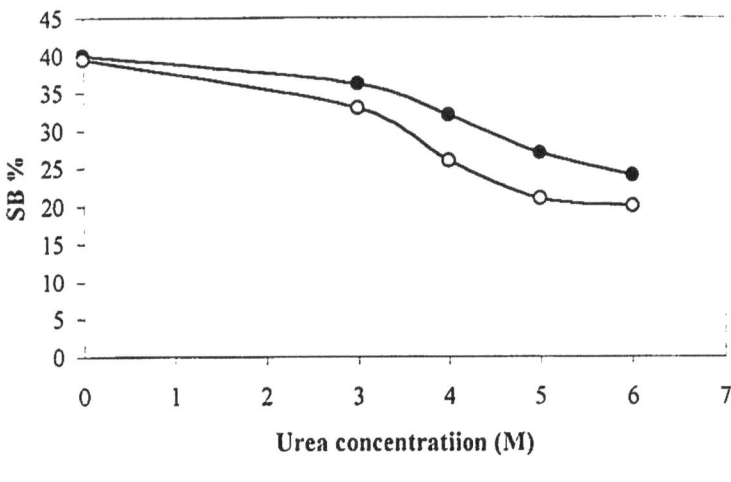

Urea concentratiion (M)

-A-

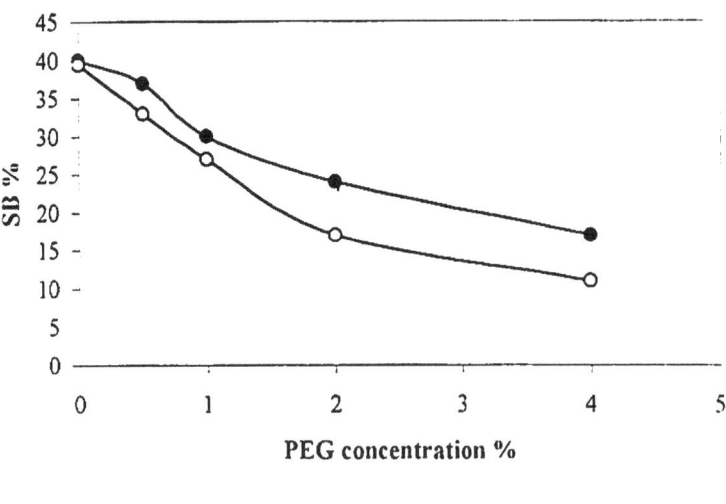

PEG concentration %

-B-

Figure (3-13): Effect of denaturating agents on the binding of lectin to erythrocyte surface glycoconjugates.

•Multiple myeloma serum sample

o Leukemia serum sample.

A: effect of urea

B: effect of PEG

Details are described in section (2.4.2.7)

3.3.3 INHIBITION STUDIES OF THE BINDING OF LECTIN TO ERYTHROCYTE SURFACE GLYCOCONJUGATES.

A number of carbohydrates were used to investigate their activities to inhibit the binding of cancerous lectin to erythrocyte surface glycoconjugates, these included sialic acid, D-glucuronic acid, fructose, mannose, and xylose. The inhibitory activity of sialic acid and D-glucuronic acid are shown in figures (3-14A) and (3-14B) respectively, table (3-8) shows the inhibitory activity of fructose, mannose and xylose.

The results revealed that the maximum inhibition of 19-23% was caused by sialic acid. D-Glucuronic acid was found to give 18-20% inhibition. Also the results obtained from this assay demonstrated that fructose, mannose and xylose have the low activities to inhibit the binding of cancerous lectin. The possible explanation for this inhibition potency is that these carbohydrates may bind to Ca^{+2} ions which are important for the completion of lectin binding to erythrocyte surface glycoconjugates. In a similar study it was found that 4-0,N-diacetylneuraminic acid and 9-0, N-diacetylneuraminic acid are the best inhibitors for a sialic acid-binding lectin from the hemolymph of Achatina fulica snail[137] also it was found that D-glucuronic acid has no inhibitory effect on the binding.

Table (3-8) Inhibition of lectin binding by fructose, mannose and xylose

Inhibitor	Concentration (mM)	Inhibition % (multiple myeloma)	Inhibition% (leukemia)
Fructose	30	2	3
Mannose	30	4	3
Xylose	30	2	-

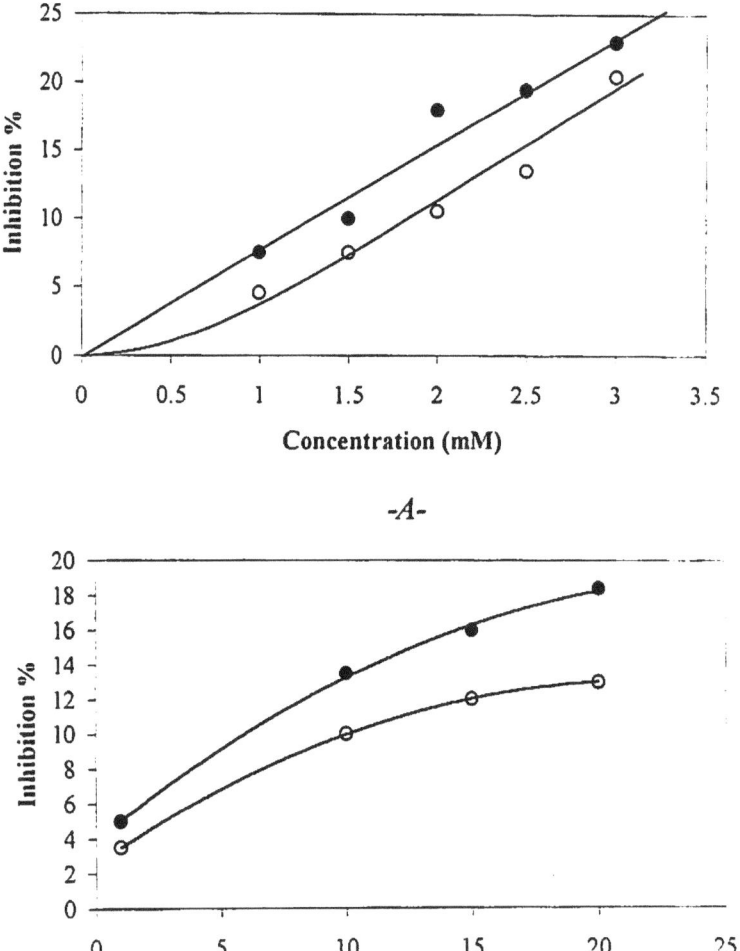

-A-

-B-

Figure (3-14): Inhibition of lectin binding to erythrocyte surface
glycoconjugates by :
-A- Sialic Acid
-B- Glucuronic Acid
Using :
•Multiple myeloma serum sample
o Leukemia serum sample.
Details are described in section (2.4.3.1), (2.4.3.2).

3.4 THE KINETIC STUDIES OF LECTIN BINDING TO ERYTHROCYTE SURFACE GLYCOCONJUGATES USING MULTIPLE MYELOMA SERUM SAMPLE

3.4.1 THE TIME-COURSE OF LECTIN BINDING TO ERYTHROCYTE SURFACE GLYCOCONJUGATES

Figure (3-15) shows the time-course of the formation of lectin-glycoconjugate complex at four different temperatures (5, 15, 25 and 37°C) in multiple myeloma serum sample. The concentration (amount) of lectin-glycoconjugate complex that formed after time (t) was calculated from the following equation:

The concentration of (lectin –glycoconjugate) complex formed after time (t) (mg/mL)	=	The concentration of lectin involved in total binding (mg/mL)	−	The concentration of lectin involved in non-specific binding (mg/mL)

- The concentration of (lectin-glycoconjugate) complex formed after time (t) was expressed in (μU), since 1μU is the concentration of lectin mg/mL which gives 40% specific binding at 37°C and 120 minutes.

The results of time-course pattern at different temperatures indicated that the lectin binding to erythrocyte surface glycoconjugates is a temperature and time dependent process, since the maximum binding can be obtained at 37°C after incubation for 120 minutes. No similar studies are available to compare our results.

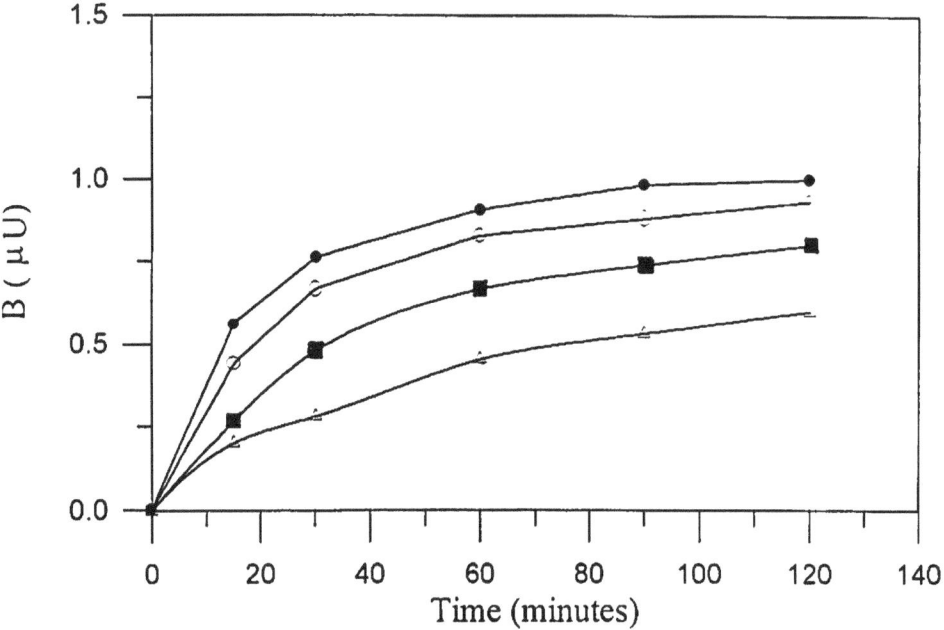

Figure (3-15): Time – course of lectin binding to erythrocyte surface

glycoconjugates at different temperatures:

(Δ) 5°C, (■) 15°C , (○) 25°C , (•) 37°C

Details are described in section (2.5.1.1).

3.4.2 DETERMINATION OF KINETIC PARAMETERS OF LECTIN BINDING TO ERYTHROCYTE SURFACE GLYCOCONJUGATES.

The time-course of lectin binding to erythrocyte surface glycoconjugates was performed to describe the kinetic parameters of the binding. The simplest proposed model representing the binding of lectin to glycoconjugates could be expressed by the following equation:

$$\text{Lectin} + G \xrightleftharpoons[K_{-1}]{K_{+1}} \text{Lectin} - G \quad \dots\dots\dots\dots\dots(1)$$

Where:

G: glycoconjugate

K_{+1}: is the rate of association of lectin with glycoconjugate.

K_{-1}: is the rate of dissociation of the complex formed under the same conditions.

At equilibrium:

$$Ka = \frac{(Lectin - G)}{(Lectin)(G)} \quad\dots\dots\dots\dots(2)$$

$$Kd = \frac{(Lectin)(G)}{(Lectin - G)} \quad\dots\dots\dots\dots(3)$$

Thus,

$$Ka = \frac{1}{Kd} = \frac{K_{+1}}{K_{-1}} \quad\dots\dots\dots\dots(4)$$

Where:

Ka: is the equilibrium constant of the association (affinity constant).

Kd: is the equilibrium constant of the dissociation of lectin-G complex.

Table (3-9) shows the values of affinity constant (Ka), Kd and maximal binding capacity (Bmax). The values of Ka and Bmax were calculated from Scatchard plot (figure (3-16)) at four different temperatures. The results obtained were similar to those obtained from Eadie-Hofstee plot (figure (3-17)). The values of Kd were calculated employing equation (4).

It is evident from the results in table (3-9) that the affinity constant (Ka) is a temperature- dependent parameter, since its value was $(13.62 \times 10^5 U^{-1})$ at 37°C while at 5°C it was $(6.6 \times 10^5 U^{-1})$, this change indicates that the lectin exhibits the maximum affinity to bind erythrocyte surface glycoconjugates at 37°C.

Table (3-9): The kinetic parameters of lectin binding to erythrocyte surface glycoconjugates Details are described in section (2.5.1.2).

Temperature °C	$Ka \times 10^5 (U^{-1})$	$Kd \times 10^{-7} (U)$	$B_{max} \times 10^{-6} (U)$
5	6.60	15.15	1.14
15	9.16	10.90	1.20
25	10.86	9.20	1.33
37	13.62	7.34	1.40

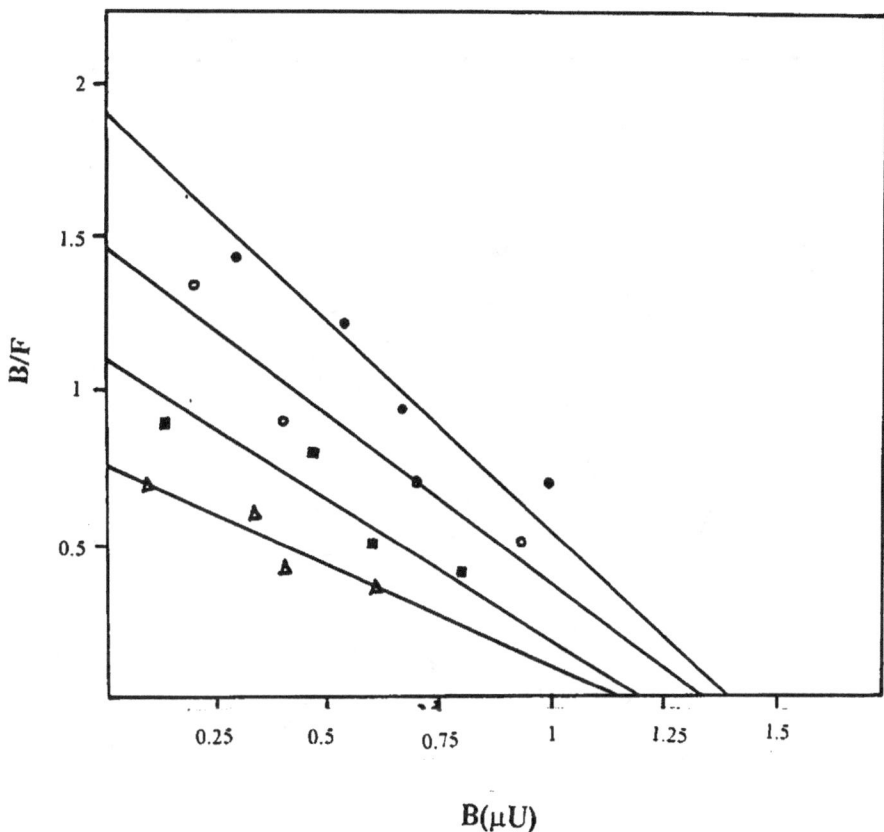

Figure (3-16): Scatchard plot of lectin binding to erythrocyte surface
glycoconjugates at different four temperatures,

(Δ) 5°C, (■) 15°C, (o) 25°C, (•) 37°C

Details are described in section (2.5.1.2).

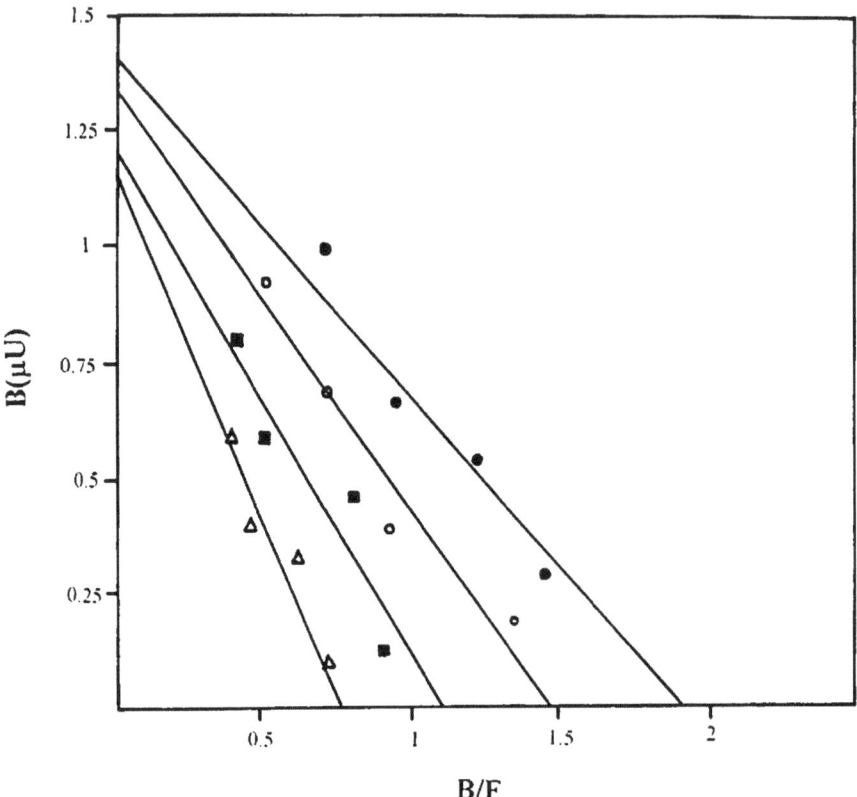

Figure (3-17): Eadie-Hofstee plot of lectin binding to erythrocyte
surface glycoconjugates at different four temperatures,

(Δ) 5°C, ($\blacksquare$) 15°C, (o) 25°C, ($\bullet$) 37°C

Details are described in section (2.5.1.2).

The time-course data obtained from figure (3-15) could be used to determine the reaction order of lectin binding to erythrocyte surface glycoconjugates using the following equation :

$$\ln(Lectin-G)_e \left[\frac{(Lectin)_t - (Lectin-G)_t(Lectin-G)_e/(G)_t}{(Lectin)_t[(Lectin-G)_e - (Lectin-G)_t]} \right] =$$

$$K_{+1}t \left[\frac{(Lectin)_t(G)_t}{(Lectin-G)_e} - (Lectin-G)_e \right] \quad \text{...............(5)}$$

Where:-

K_{+1}: is the kinetic association constant in $U^{-1}.min^{-1}$.

(Lectin)$_t$: is the concentration of lectin at time (t).

(G)$_t$: is the concentration of glycoconjugate at time (t).

(Lectin-G)$_t$: is the concentration of lectin-glycoconjugate complex at time (t).

(Lectin-G)$_e$: is the concentration of lectin-glycoconjugate complex at equilibrium.

Equation (5) represents the second order kinetics. But in our work in which the percent of specific binding was, in some cases, small and most of the lectin remains free and only small fraction binds even at equilibrium, i.e. (Lectin)t>>(Lectin-G)$_e$ thus,

$$(Lectin)_t >> \frac{(Lectin-G)_t(Lectin-G)_e}{(G)_t} \ and \ \frac{(Lectin)_t(G)_t}{(Lectin-G)_e} >> (Lectin-G)_e$$

So that the following equation could be used in order to fit the pseudo first order kinetics:

$$\ln \frac{(Lectin-G)_e}{(Lectin-G)_e - [Lectin-G]_t} = K_{+1}t \frac{(Lectin)_t(G)_t}{(Lectin-G)_e} \quad \text{...............(6)}$$

Figure (3-18) shows the plot of $\ln \dfrac{(Lectin-G)e}{(Lectin-G)e-(Lectin-G)t}$ against time (t), the slope of the resulting straight line equal to the observed value of first-order rate constant in min^{-1}. The rate constant (K_{+1}) was calculated at four different temperatures by using the following equation:

$$K_{obs} = K_{-1}\dfrac{(Lectin)(G)t}{(Lectin-G)e}$$

$$\therefore K_{obs} = K_{-1}(Lectin)t \dots\dots\dots\dots\dots\dots\dots\dots\dots\dots\dots\dots\dots(7)$$

The value of K_{-1} at four different temperatures were calculated by using equation (4). Also the half life time of association $(t1/2)_{ass.}$, which represents the time needed for the formation of half amount of the complex at equilibrium, was determined from the concentration of the complex at equilibrium and the time-course curve. While the half life time of dissociation $(t1/2)_{diss.}$ was calculated from:

$$(t_{1/2})_{diss} = \ln\dfrac{2}{K_{-1}} = \dfrac{0.693}{K_{-1}}\dots\dots\dots\dots\dots\dots\dots\dots\dots(8)$$

The values of $K_{obs.}$, K_{+1} , K, $(t1/2)_{ass.}$ and $(t/1/2)_{diss.}$ at four different temperatures were summarized in table(3-10). The results revealed that the association rate constant (K_{+1})at $37°C$ is the highest one among other values at 5, 15 and $25°C$. Also the results show that the dissociation rate constant (K_{-1}) changes with temperature this indicates that the rate of dissociation of lectin – glycoconjugate complex depends on temperature.

Table (3-10): The effect of temperature on the kinetic parameters of lectin binding to erythrocyte surface glycoconjugates. Details are described in section (2.5.1.1).

Temperature °C	K_{obs} (min.$^{-1}$)	$K_{+1} \times 10^3$ (U.min)$^{-1}$	$K_1 \times 10^3$ (min^{-1})	$(t_{1/2})_{ass.}$ (min.)	$(t_{1/2})_{diss.}$ (min)
5	0.023	9.2	13.93	28.5	49.75
15	0.028	11.2	12.21	23.5	56.76
25	0.036	14.4	13.25	15.0	52.31
37	0.0455	18.2	13.36	10.5	51.88

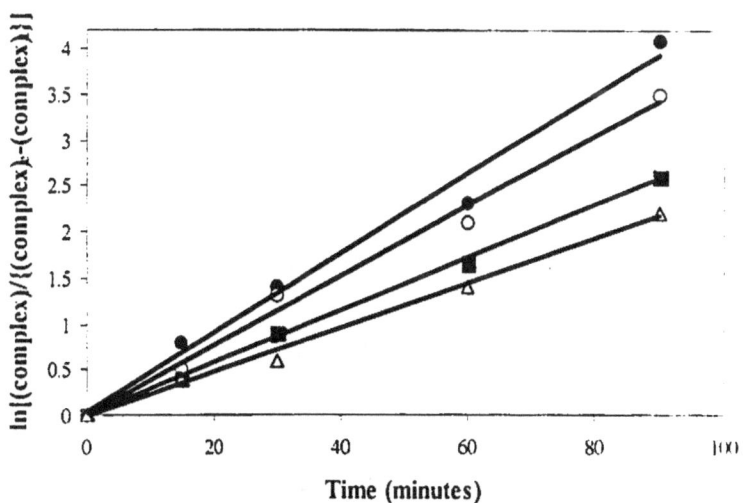

Figure (3-18): Kinetics of lectin binding to erythrocyte surface glycoconjugates at four temperatures,

(Δ) 5°C, (■) 15°C, (o) 25°C, (•) 37°C

Details are described in section (2.5.1.1).

3.4.3 DETERMINATION OF HILL COEFFICIENT (n)OF LECTIN BINDING TO ERYTHROCYTE SURFACE GLYCOCONJUGATES

Using the data obtained from Scatchard analysis, the values of

$$\log \left(\frac{B}{B_{max} - B} \right) \quad \text{against log F were plotted according}$$

to Hill equation, figure (3-19). The value of Hill-coefficient (n) equals the slope of the resulting straight line. The values of Hill-coefficient (n) were obtained at 5,15,25 and 37°C they were 1.22, 1.22, 1.26 and 1.28 respectively. The cooperativity of the lectin binding sites could be estimated through the determination of Hill- coefficient (n). The results obtained in this work indicates that the cooperativity of lectin binding sites was low affected by temperatures.

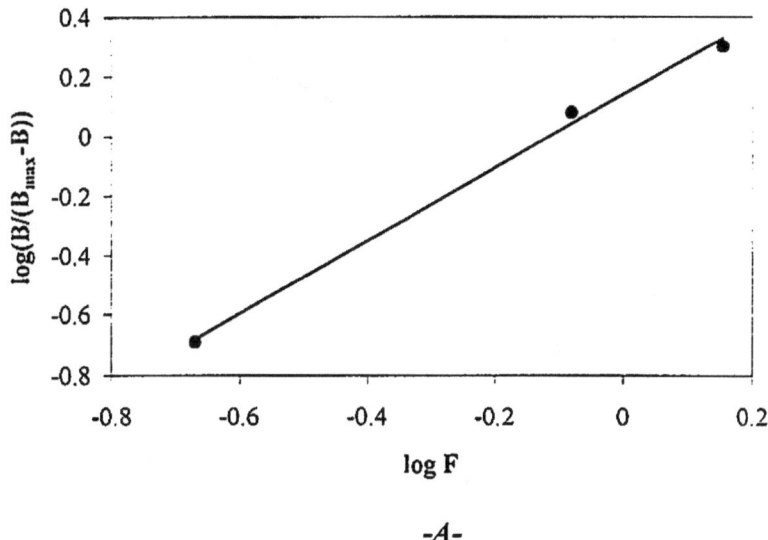

-A-

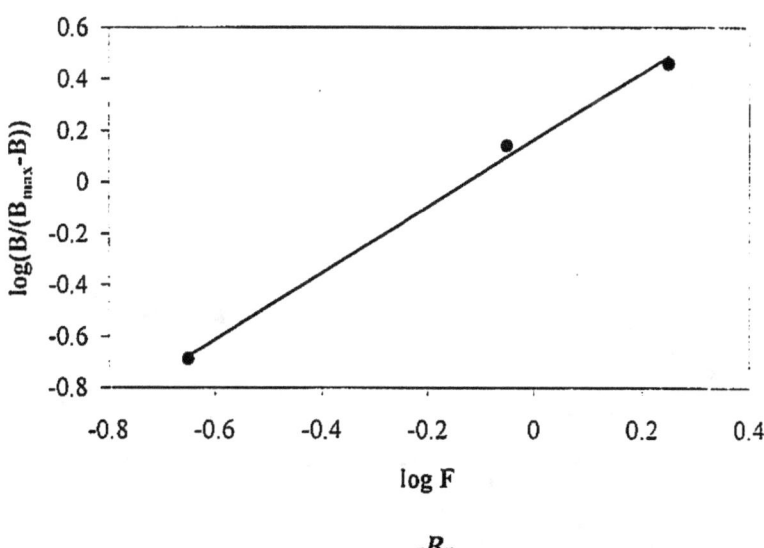

-B-

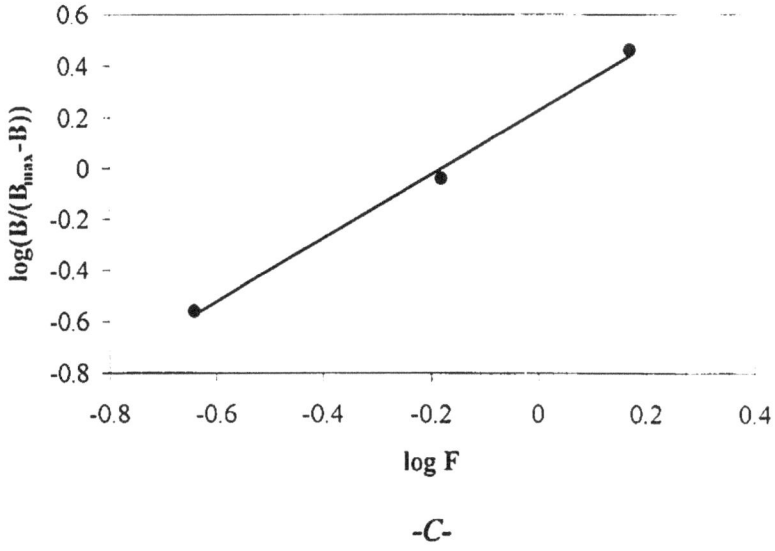

log F

-C-

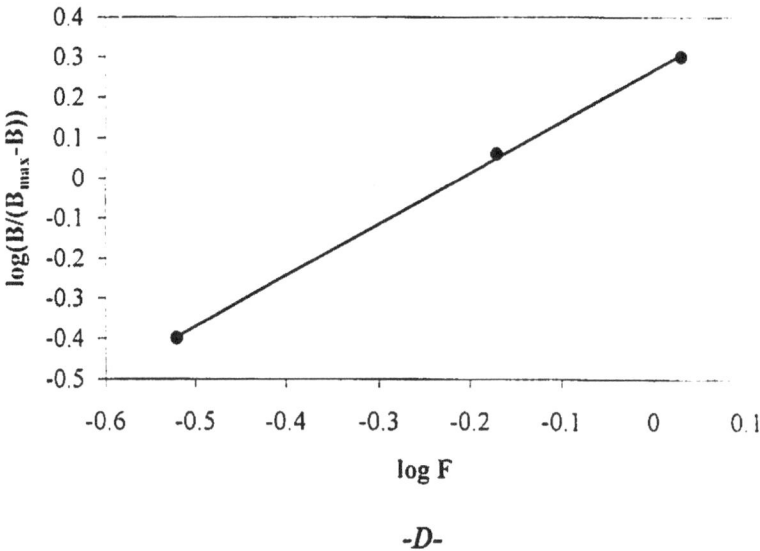

log F

-D-

Figure (3-19): Determination of Hill-coefficient (n) of lectin binding to erythrocyte surface glycoconjugates at four temperatures,

-A- 5°C, -B- 15°C, -C- 25°C, -D- 37°C

Details are described in section (2.5.1.3).

3.5 THE THERMODYNAMICS OF THE LECTIN BINDING TO ERYTHROCYTE SURFACE GLYCOCONJUGATES

1- Thermodynamic parameters of standard state

The dependence of the equilibrium binding constant (affinity constant)for the binding of lectin to erythrocyte surface glycoconjugates on temperature can be observed from Van't Hoff plot (figure (3-20)).

The results obtained from Van't Hoff plot revealed that ΔH° in general had a positive value of 17.36 KJ/mol., and that the reactions were nearly endothermic. The small positive value of ΔH° may indicate a favorable interaction between the lectin and glycoconjugate subgroups. These include the non-covalent interaction which are fundamentally electrostatic in nature such as charge-charge, charge-dipole, dipole-dipole, charge-induced dipole, dipole-induced dipole interactions, and hydrogen bonds. The sum of these types of interactions can yield some stabilization to the folded structure of the complex.

Table (3-11) shows the values of ΔG° at four temperatures (5,15,25,and 37°C). The results revealed that the ΔG° values increases with decreasing temperature, since its value was -30.97 KJ/mol. at 5°C, -32.85 KJ/mol. at 15°C, -34.43 KJ/mol. at 25°C and -36.39 KJ/mol. at 37°C, it can be said that the lectin binding to erythrocyte surface glycoconjugates needs higher energy at low temperatures. Also the negative values of ΔG° indicates the stability of lectin-glycoconjugate complex, subsequently the high affinity of the reactant. The high negative values of ΔG° indicates that the binding of lectin to glycoconjugates is a spontaneous reaction. In addition, these values are controlled by high positive ΔS° values, table(3-11). The results show that the values of ΔS° decrease with increasing temperature, since its value reaches 173.38J/mol.K at 37°C , this can be attributed to the more stable and more arranged status of lectin-glycoconjugate complex at 37°C.A high values of positive ΔS° suggest that the binding spontaneity was entropically driven. Entropy was the driven force for the occurrence of the binding, this indicates that the hydrophobic interactions played an important role in stabilizing the complex.[138]

Table (3-11): Thermodynamic parameters at standard state of lectin binding to erythrocyte surface glycoconjugates. Details are described in section (2.5.2).

Temperature °C	ΔH° (KJ/mol.)	ΔG° (KJ/mol.)	ΔS° (J/mol.K)
5	17.36	-30.97	173.84
15	17.36	-32.85	174.34
25	17.36	-34.43	173.79
37	17.36	-36.39	173.38

2- Thermodynamic parameters of transition state

The transition state theory proposes that the association of two substances to form the final product proceeds through the formation of an activated complex(transition state). Consequently, the interaction of lectin with erythrocyte surface glycoconjugates can be represented as follows:

Lectin - G ⟶ (Lectin –G) ⟶ Lectin G

(glycoconjugate)   an activated complex (transition state)   (The final product)

The thermodynamic parameters of the transition state (ΔH^*, ΔG^*, and ΔS^*) could be calculated employing Arrhenius equation and the kinetic constant.

Figure (3-21) shows Arrhenius plot of $\ln K_{+1}$ against $1/T$ values. The slope of the straight line represents the activation energy (Ea).

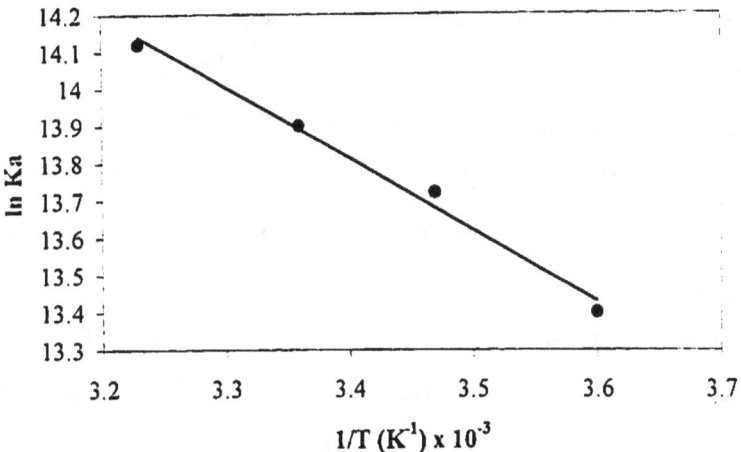

Figure (3-20): Van't Hoff plot for the binding of lectin to erythrocyte surface glycoconjugates. Details are described in section (2.5.2).

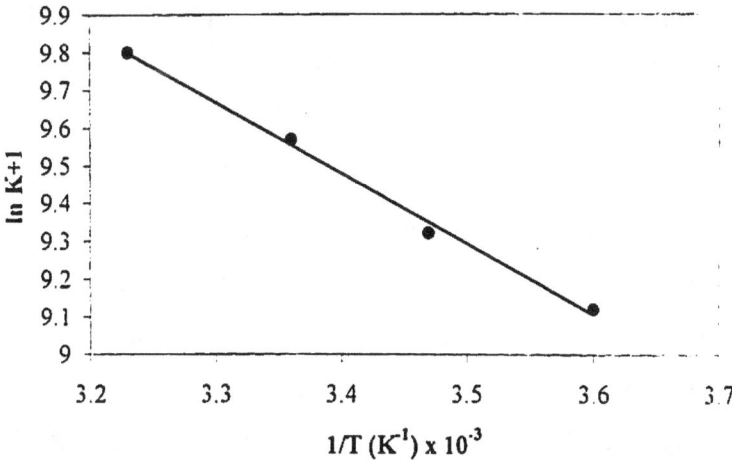

Figure (3-21): Arrhenius plot for the binding of lectin to erythrocyte surface glycoconjugates. Details are described in section (2.5.2).

Table (3-12) shows the values of thermodynamic parameters of the transition state (Ea, ΔH^*, ΔG^*, and ΔS^*). The high value of activation energy (15.27KJ/mol.) represents the required energy to overcome the energy barrier of the transition state for the formation of lectin-glycoconjugate complex.

Also the value of activation energy is in accordance with the high positive values of ΔG^* which indicates that the formation of the activated complex is a non-spontaneous process.

The results show that the values of ΔH^* decrease with increasing temperature, since its value was 12.95 KJ/mol. at 5°C, 12.87KJ/mol. at 15°C, 12.79 KJ/mol. at 25°C and 12.69 KJ/mol. at 37°C. The slight changes in the values of ΔH^* at different temperatures could be attributed to the dependence on ΔH^* an activation energy (Ea) through the equation:

$$\Delta H^* = Ea - RT$$

Since the numerical value of RT is too small in comparison with the value of activation energy for the binding of lectin to glycoconjugates.

Table (3-12) shows the increasing of ΔS^* values with the elevation of temperature ΔS^* values were -121.83 J/mol.K at 5°C, -122.37 J/mol.K at 15°C, -122.38 J/mol.K at 25°C and -122.74 J/mol.K at 37°C.

The high negative values of ΔS^* indicate that the activated complex (lectin-G) involved in the binding process had a more arranged structure than the starting reactants (lectin and glycoconjugate).

From the results obtained for the thermodynamic parameters in the transition state, it can be concluded that the positive values of ΔH^* and high positive values of ΔG^* are favorable to overcome the energy barrier of the transition state, the high negative values of ΔS^* mean more arranged structure for the activated complex.

The values of the thermodynamic parameters obtained from the study of lectin binding to erythrocyte surface glycoconjugates, give a distinct idea about the nature of forces that regulate the formation of lectin-glycoconjugate complex. The thermodynamic model describing the formation of the complex was suggested using the thermodynamic parameters of both the standard and the transition states. The model is illustrated in figure (3-22).

The thermodynamic model proposes that the formation of lectin-glycoconjugate complex undergoes three thermodynamic states.

The thermodynamic state A represents the initial energy level of lectin and glycoconjugate (G). The thermodynamic state B represents the association of the two species to form the activated complex (Lectin-G). At thermodynamic state C, complete binding of lectin with glycoconjugate (formation of the lectin-glycoconjugate complex).

The model involves two steps, the reaction at step1 is associated with positive ΔG^* value, this indicates that the binding of lectin to glycoconjugates in this step requires external energy. Also in step 1, the lectin binding shows negative value for entropy change (ΔS^*),this negativity indicates the alteration in the structure of lectin-glycoconjugate transition complex to a more arranged one. At step2, the contribution of the activated complex in more interactions, giving the fully interacting complex (Lectin-G).

It is proposed that the formation of a protein-ligand complex occurs in two steps, the first is the stabilization of the complex by hydrophobic interactions and the second is the stabilization by short range interactions, such as electrostatic interactions, hydrogen bonding and Van der Waals interactions [139].

Hydrophobic interactions contribute to the complex stability via high positive entropy change ($\Delta S^\circ > 0$), while the electrostatic interactions, hydrogen bonding, and Van der Waals interactions contribute to the stability of the lectin-glycoconjugate complex via negative entropy change ($\Delta S^\circ < 0$) [139-140].

The thermodynamic data from the present study indicate that the binding of lectin to erythrocyte surface glycoconjugates is entropy driven and is in agreement with the concept that hydrophobic interactions play an important role in such reactions.

Table (3-12): Thermodynamic parameters at transition state of lectin binding to erythrocyte surface glycoconjugates. Details are described in section (2.5.2).

Temperature °C	Ea (KJ/mol.)	ΔH^* (KJ/mol.)	ΔG^* (KJ/mol.)	ΔS^* (J/mol.K)
5	15.27	12.95	46.83	-121.83
15	15.27	12.87	48.12	-122.37
25	15.27	12.79	49.26	-122.38
37	15.27	12.69	50.74	-122.74

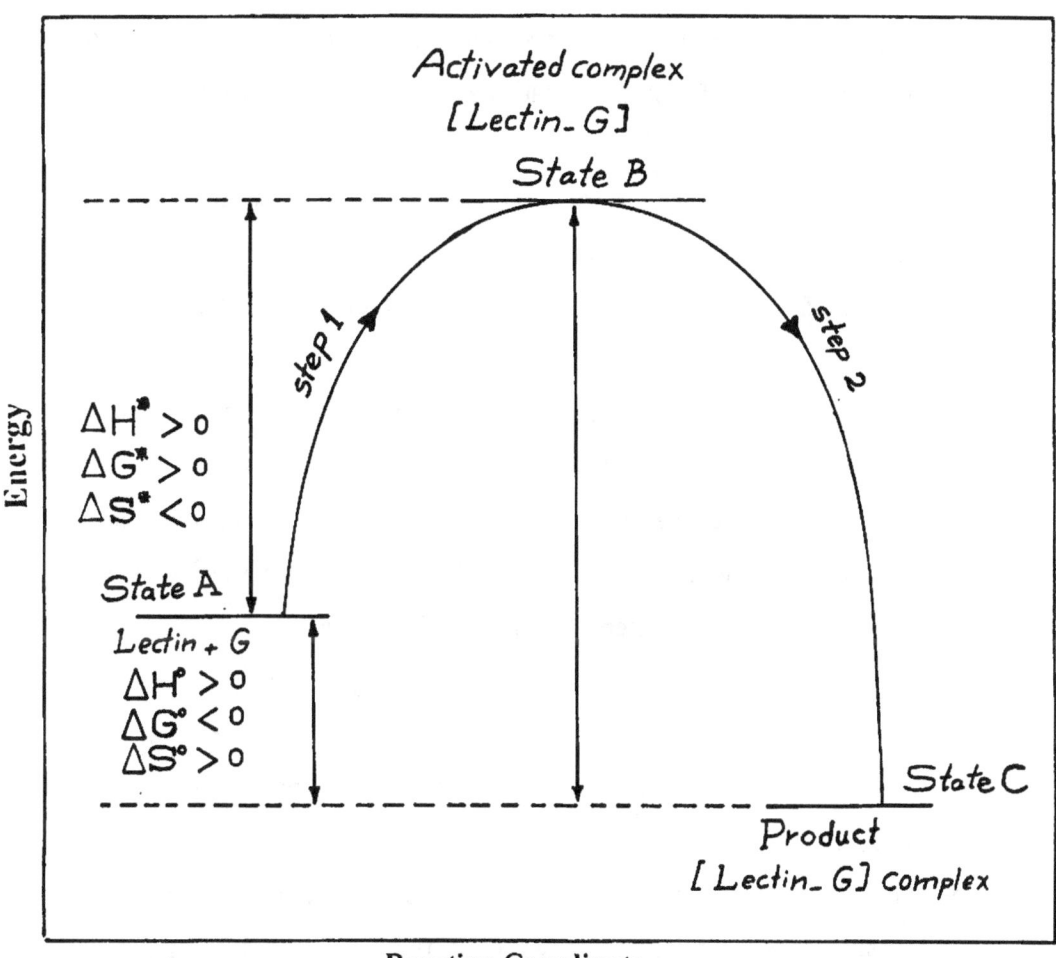

Figure (3-22): General energy diagram and thermodynamic model
applied to the complex formation between lectin and erythrocyte
surface glycoconjugates.

Conclusions

1- From the present study it can be concluded that the reported elevations in the levels of total and lipid-bound sialic acid could be used to differentiate patients with cancer from healthy individuals

2- The decreased values of SOD activity in cancer patients negatively correlate with high LBSA levels, the index LBSA/SOD may be clinically important tool for the diagnosis of cancer diseases.

3- The study of lectin binding to human erythrocyte surface glycoconjugates may prove a valuable tool to investigate the localization, characterization and assessment of glycoconjugates. the binding reaction could be used to characterize the cancerous lectins through the different patterns of binding.

4- The kinetic study of lectin binding to glycoconjugates revealed that the binding reaction is a time and temperature dependent process. The results showed that the reaction is pseudo first order at (5, 15, 25 and 37°c)

5- The thermodynamic study and its parameters could be utilized to investigate the relation between lectin and glycoconjugates in vitro. Lastly, since the driving force for the formation of (lectin-glycoconjugate) complex is solely thermodynamic ($\Delta S''>o$), other multivalent ligand-macromolecular systems must possess the same properties.

Future work

1- Measurement of TSA and LBSA in sera of patients pre and post chemotherapy treatment to assess its activity.

2- Combination of TSA with other markers to improve the specificity of TSA as tool for the detection of human cancer

3- Purification of the lectin from sera of patients with multiple myeloma and leukemia

4- Purification of the lectin from the bone marrow of patients with multiple myeloma and leukemia

5- Study the ability of these purified lectins to agglutinate other types of human erythrocytes.

6- To examine the surface component of cancerous cells, it is prefarable to study the ability of purified lectins to bind cancerous cells related to different stages of the disease.

References

1-Ajit Varki and Sandra Diaz, *J. Biol. Chem.* (1983) 258 (20): 12465 – 12471.

2-Katsuko Anazawa, etal. *Chem. Pharm. Bull.* (1988) 36 (12): 4976 – 4979.

3-Sheshadri Narayanan. *Ann. Clin. Lab. Sci.* (1994) 24 (4) 376 – 384

4-Edelman, G.M. *Science* (1983) 219: 450 – 451.

5-Miguel A. Ferrero, etal. *Biochem. J.* (1996) 317: 157 – 165.

6-Schauer, *R. Adv. Carbohydr. Chem. Biochem* (1982) 40: 131 – 234.

7-Aldo Lagana, etal. *Clin C him. Acta.* (1995) 243: 165 – 179.

8-Surggak Kim, and Kyu Yang Yi. *J. Org. Chem* (1986) 51: 2615 – 2617.

9-Mepur H. Ravindranath., etal. J.Biol. Chem. (1985) 260 (15): 8850 – 8856.

10-Albert L. Lehninger, *Biochemistry*. 2nd . ed. Worth Publishers, Inc. 1975 pp 651.

11-Cornforth, J.W., etal. *Bio chem. J.* 1958, 68, 57.

12-Roseman, S. and Comb, D.G. *J. Amer. Chem. Soc.* 1958, 80, 3166.

13-Carroll, P.M. and Cornforth, G.W. *Biochem. Biophys. Acta.* (1960) 39,161.

14-Heinz – Werner H. and Reinhard Brossmer. *Helv. Chem. Acta.* (1986) 69: 2127 – 2132.

15-Miburn, J. J. *Biol. Chem.* (1989) 264,483.

16-Hakomori, S. *Adv. Cancer. Res.* (1974) 18: 265 – 315.

17-Lipton, A, Harvey HA, Delong SA, etal. Cancer (1979) 43: 1766 – 1771.

18-Corfield, A.P. and Schauer, R. (1982) In Sialic Acids: Chem., Metabol and Funchion, cell biol. Monographs Vol. 10 pp 3-50 Springer Verlag, Wien.

19-Svennerbolm, L. *Biochim. Biophys. Acta.* (1957) 24: 604 – 611.

20-Warren, L. *J. Biol. Chem.* (1959) 234: 1971 – 1975.

21-Shukla, AK. And Schauer, R. *Hoppe – Seyler's Z physiol. Chem.* (1981) 362: 236 – 237.

22-Shamberger, R.J. *Anticancer. Res.* (1986) 6: 717 – 720.

23-Hara, S., Takemori, Y. etal *Anal. Biochem.* (1987) 164: 138 – 145.

24-Schuhr, EMJ. etal. *Tumor Biol.* (1992) 13: 121 – 132.

25-Svennerblom, L. and Freedman, RA. *Biochim. Biophys. Acta* 1980 617: 97 – 109.

26-Samuel J. Danishefsky, Michael P. Deninno, and Shu – huichen. *J. Am. Chem. Soc.* (1988) 110, 3929 – 3940.

27-Marikovsky, Y., Ben – Bassat, H., etal. J. Natl. Cancer. Inst. (1979) 62: 285 – 292.

28-Crook, M. *Clin. Biochem.* (1993) 26: 31 – 38.

29-Schauer, R. *Trends Biochem. Sci.* (1985) 10: 357 – 360.

30-Marion, E. Reid, David. J. *Anstee, etal. Biochem J.* (1987) 244: 123 – 128.

31-Cheresh, DA., Varki AP, etal. *Biochem. Biophys. Res. Commun.* (1986) 72: 456 – 461.

32-Varki, A., and Kornfeld, S., *J. exp. Med.* (1980) 152: 532 – 544.

33-Fukushima, K. *J. exp. Med.* (1991) 163: 17 – 30.

34-Itai, S., Arii, S., etal. *Cancer* (1988) 61: 775 – 787.

35-Hakomori, S. *Cancer. Res.* (1985) 45: 2405 – 2414.

36-Ichiroh Shemada, etal. *J. Gasteroenterol.* (1995) 30: 21 – 27.

37-Obzen, T. *Ann. Clin. Biochem.* (1991) 28: 44 – 48.

38-Toshihiko Murayama, etal. *Int. J. Cancer* (1997) 70: 575 – 581.

39-Prabhodas, S. Patel, etal. *Neoplasma* (1995) 42 (5): 271 – 276.

40-Oztokatli, A. etal. *Int. Urol. Nephrol.* (1992) 24: 125 – 129.

41-Yue, K. etal. *Acta. Acad. Med. Sci.* (1995) 17 (2), 128 – 132.

42-Patel, P.S., Rawal, G.N., and. Balar, D.B. *Gynecol. Oncol.* (1993) 50 (3): 294 – 300.

43-Gatchev, O., Rastam, L. etal. *Br. J. Cancer* (1993) 68: 425 – 427.

44-Vadralova, E. and Borovansky, J. *Cancer. Lett.* (1994) 78: 171 – 172.

45-Riley, M., Tautu, C., Verazin, G. *Clin. Chem.* (1990) 36: 161 – 163.

46-Silver, H.K.B., Rangel, D.M. and Morton, D.L. *Cancer* (1978) 41: 1497 – 1499.

47-Plucinsky, M.C., etal. *Cancer* (1986) 58: 2680 – 2685.

48-Polivkova, J., Vosmikova, K. and Hora, K.L. *Neoplasma* (1992) 36 (4): 233 – 236.

49-Kiricuta, O., Bojan, O.,Comes, R. and Christian, R. *Arch. Geschwulst Forsch.* (1979), 49: 106 – 112.

50-Robert, S. Bresalier, etal. *Cancer Res.* (1990) 50: 1299 – 1307.

51-Peter Altevogt, etal. *Cancer Res.* (1983) 43: 5138 – 5144.

52-J. Friedman, H. Levinsky, D. Allalouf, and A. Staroselsky, *Cancer Lett.* (1988) 43: 79 – 84.

53-Yogeeswaran, G. and Salk, P.L., *Science* (1981) 212: 1514 – 1516.

54-Stringou, E., Chondros, K., etal, *Anticancer Res.* (1992) 12: 251 - 255.

55-Schuttr, EMJ., etal. *Tumor. Biol.* (1992) 13:121 – 132.

56-Patel, P.S. Baxi, B.R. and. Balar, D.B. *Neoplasma* (1989), 36 (1): 53 – 59.

57-Albert, L. Lehninger, *Biochemistry*, 2nd .ed. Worth publishers., Inc. (1975) pp273 – 274.

58-Lubert Stryer, *Biochemistry* 3rd. ed W.H. Freemen./ New York (1988) pp298.

59-Lopez Saez. J.B, Senra' – Varela, A. *Int. J. Biol. Mark.* (1995) 10 (3): 174 – 179.

60-Roos, F. *Biochem. Biophys. Acta.* (1984) 738: 263 – 284.

61-Bhuvarahamurthy, V. etal. *Int. J. Gynecol. Obst.* (1995) 48: 49 – 54.

62-Larry, W. Oberley and Garry, R. Buettner *Cancer Res.* (1979) 39: 1141 – 1149.

63-Misra, HP. and Frido vich, I. *J.Biol. Chem.* (1972) 247: 6960 – 6962.

64-McCord, JM. and Fridovich, I., *J. Biol. Chem.* (1969) 244: 6049 – 6055.

65-Knee, JK. and Mitidieri, E. and Affonso, OR, *Cancer. Lett.* (1991) 57: 199 – 202.

66-Maneva, A., Michailova, D. etal., *Eur. J. Cancer Prevention* (1995) 4: 429 – 435.

67-Dipti, G. and Fred, C. Brewer, J. *Biochemistry* (1994) 33: 5526 – 5530.

68-Irwin J. Goldstien, etal. *Nature* (1980) 285: 66.

69-Barondes, S.H., *Ann. Rev. Biochem.* (1981) 50: 207 – 231.

70-Goldstien, I. J. and Hayes, C.E., *Adv. Carbohydr. Chem. Biochem* (1978) 35: 127 – 340.

71-Barondes, S.H. *Science* (1984) 223: 1259 – 1264.

72-Lis,H. and Sharon, N. *Ann. Rev. Biochem.* (1986) 55: 35 – 67.

73-Yamamoto, K., etal. *Biochem. J.* (1981) 195: 701 – 713.

74-Cummings, R.D. and Kornfeld, S., *J. Biol: Chem.* (1982) 257: 11235 – 11240.

75-Tollefesen, S.E. and Kornfeld, S., *J. Biol. Chem.* (1983) 258: 5165 – 5171.

76-Peters, B.P. etal. *Biochemistry* (1979)) 18: 5505 – 5511.

77-Roche, A.C. etal. *FEBS lett* (1975) 57, 245 – 249.

78-Mohan, etal. *Biochem. J.* (1982) 203: 253 – 261.

79-Miller, R.L. J. *Invertebr. Pathol.* (1982) 39: 210 – 214.

80-Babal, P. *Biochem. J.* (1994) 299 (2): 341.

81-Kawagishi, *H. FEBS Lett.* (1994) 340: 56.

82- Monsigny, M. and Kieda, C. *Biol. Cell* (1983) 47: 95 – 110.

83-Hasilik, A. and Vonfigura, K., *Ann. Rev. Biochem.* (1986) 55.

84-Raz, A. and Lotan, R., *Cancer Res.* (1981) 41: 3642 – 3647.

85-Monsigny, M., Roche, A.C. and Midoux, P. *Biol. Cell.* (1984) 51: 187 – 196.

86-Ashwell, G. and Harford, J., *Ann. Rev. Biochem.* (1982) 51: 531 – 554.

87-Levi, G. and Teichberg, V. I, *J. Biol. Chem.* (1981) 256: 5735 – 5740.

88-Beyer, E.C., Zweig, S.E. and Barondes, S.H. *J. Biol. Chem.* (1980) 255: 4236.

89-Springer, G.F. and Desai, P.P., *Biochemistry* (1971), 10: 3749.

90-Cohen, E.(1984) *Recognition Proteins, Receptors and Probes*: invertebrates progress in clinical and biological Research vol. 157 Newyork; lis pp 207.

91-Gilbsa – Garber, N. etal. *FEBS Lett* (1985) 81: 267 – 270.

92-Vasta, G.R., Cheng, T.C. Marchalonic, *J.J Cell Immunol.* (1984) 88: 475 – 488.

93-Bohlool, B.B. and Schmidt, E.L. *Science* (1974) 185: 269 – 271.

94-Mirelman, D., etal. *Nature* (1975) 256: 414 – 416.

95-Mishkind, M., etal. *J. cell. Biol.* (1982) 92: 753 – 764.

96-Regoeczi, E. etal *Proc. Nath. Acad. Sci. USA* (1982) 79: 2226.

97-Stahl, P.D. etal, *Biol.cell* (1984) 51: 215 – 218.

98-Sahagian, G.G. *Biol.cell* (1984) 51: 207 – 214.

99-Lotan, R., Lotan, D., Raz, A., *Cancer Res.* (1985) 45: 4349 – 4353.

100-Barondes, S.H., etal. *Biol.cell.* (1984) 51: 165 – 172.

101-Hoffbrand, A.V. and Pettit J.E. Essiental haematology2nd ed Blackwell Scientific publications 1989 pp. 173 - 181.

102-Lubert Stryer, *Biochemistry*, 3rd.ed. W.H.Freeman / New.York. (1988) pp 895.

103-Christopher R.W. Edwards and Ian, A.D. Bouchier, *Davidson's Prinsiples and Practice of Medicine* (1992) 16th. Ed Churchill Livigstone.

104-Giuseppe, C., Giuseppe, A. etal. *Brit. J. Haematol.* (1990), 75: 373 – 377.

105-Kaiser,U., Auerbach, B. and Oldenburg, M. *Leuk. Lymphoma* (1996) 20: 389 – 395.

106-Merlini, G., Perfetti, V. etal. *Brit.J. Haematol.* (1993) 83 (4): 595 – 601.

107-Moriyama, T.,Tozawa, T., etal. *J. Chromatogr.* (1991), 571: 61 – 72.

108-Moriyama, T., Tozawa, T. etal. *Clin. Chim. Acta.* (1995), 223: 127 – 134.

109-Cohen, A.M., Allalouf, D., etal *Eur. J. Haematol.* (1989), 42: 289 – 292.

110-Crook, M.A., Couchman, as.and Tutt, P. Brit. *J. Biomed. Sci.* (1996) 53: 185 – 186.

111-Lowry, OH., Rosebrough, NJ. Farral, AL,etal *J.Biol.Chem.* (1951) 93:265.

112-Katopodis, N. etal *Cancer. Res.* (1982) 42: 5270 – 5275.

113-Weimer and Mashin, *Clin. Chem. Prinsiple and technics.*

114-Winterbourn, etal. *J. Lab. Clin Med.* (1975) 85 (2): 337 – 341.

115-Miller, etal. *Methods in Enzymology* (1987), Vol 138: pp 527 – 530. Academic press. NewYork.

116-Liener, I. *Arch. Biochem. Biophys.* (1955). 54: 223.

117-Lis, H. and Sharon,N. *Methods in Enzymology* (1972) Vol. 28 pp 360.

118-Scatchard, G. *Ann. NY. Acad. Sci* (1949) 51: 660.

119-Sherblow, A.P. etal. *J. Biol.Chem.* (1980) 255: 783.

120-Gail, H.M., etal, *JNCI* (1986), 76: 805.

121-Horgan, I.E. *Clin. Chem Acta.* (1982), 118: 327.

122-Erbil, etal, *Cancer* (1986) 57: 1889.

123-Dimisttian. AM. Etal. *Cancer* (1982), 50: 1815 – 1819.

124-Erbil, K., Jones, J. and Klee,G. *Cancer* (1985) 55: 404 – 409.

125-Black, P.H. *N. Engl. J. Med.* (1983) 303: 1415 – 1416.

126-Dwivedi, C. etal, *Experientia* (1990) 46: 91 – 94.

127-Dnistrian, M. and Schwartz, K. *Clin Chem*. (1981) 27 (10): 1737 – 1739.

128-Vivas. I. etal *Gynecol. Oncol*. (1992) 46 (2): 157 – 162.

129-Toumbis, etal. *Anticancer. Res*. (1992) 12 (4): 1267 – 1270.

130-Voigtmann, R. Pokorny, J. and Meinshausen, A. *Cancer* (1989) 64: 2279 – 2283.

131-Mrochek J.E. etal, *Clin.Chem*. (1976) 22: 1516 – 1524.

132-Patel, P.S. etal. *Cancer Lett*. (1990) 51: 203 – 208.

133-Breadly, WP. etal, *Cancer* (1977) 40: 2264 – 2272.

134-Fernandez – pol, J.A. etal, *Cancer Res* (1987) 42: 609 – 617.

135-Oberly, L., Oberly. T. *Mol. Cell Biochem*. (1988) 84: 147 – 153.

136-Kaplan, A. (1998) Clinical Chemistry, theory, Analysis and corelation, 2nd . edition. pp 180

137-Chitra, M. and Sujata, B., *Biochem. Biophys. Res Commun*. (1987) 148 (2): 795 – 801.

138-Waelbroeck, M. and Van Obeerghen, E. and Demeyts, *P.J.Biol. Chem*. (1979), 254: 7736.

139-Blumenthar, D.K. and Stull, J.T. *Biochemistry*. (1982) 21: 2386.

140-Laport, DC. And Wierman, EM. nd Storm, Dl. *Biochemistry* (1980), 19: 3814.

www.ingramcontent.com/pod-product-compliance
Lightning Source LLC
Chambersburg PA
CBHW080813180526
45168CB00006B/2434